AF601557

Biology and Culture Potential of *Chitala chitala*

The Authors

Dr. Anisa Mitra completed her M.Phil. in Fisheries Management from University of Calcutta and obtained her Ph.D. from Department of Zoology, Fisheries and Aquaculture Division, University of Calcutta. She started her career as a Lecturer (Zoology) in St. Paul's Cathedral Mission College, Kolkata. Presently, Dr. Mitra is working as Lecturer in Zoology, New Alipore College, University of Calcutta.

Dr. Pratap Kumar Mukhopadhyay obtained Ph.D. from University of Calcutta. He was a Post doctoral fellow at the Institute of Aquaculture, University of Stirling, U.K. and served as a Visiting Scientist at Unite de Recherchesen Hydrobiologie (INRA) St-PEE-SUR-Nivelle,France. He worked as a Senior Scientist at the International Rice Research Institute and served as FAO Consultant in Aquaculture in Kenya and Uganda. He retired as a Principal Scientist from ICAR-CIFA, Regional Centre, Rahara, Kolkata.

Prof. (Dr.) Sumit Homechaudhuri did his M.Phil. in Environmental Science and subsequently received his Ph.D. in Fisheries Science under University of Calcutta. He was a Gold medallist of University of Calcutta. He started his career as a Lecturer in Zoology, Darjeeling Govt. College, India and subsequently became an Assistant professor and a Reader in the erstwhile Presidency College. Presently, Dr. Homechaudhuri is working as Senior Professor in the Department of Zoology in University of Calcutta. He acted as the former Head and Ph.D. Convenor of Zoology Department, University of Calcutta.

Biology and Culture Potential of *Chitala chitala*

Anisa Mitra
Ph.D., M.Phil.,
Lecturer in Zoology New Alipore College,
University of Calcutta

Pratap Kumar Mukhopadhyay
Ph.D., Retired Principal Scientist, ICAR- CIFA

Sumit Homechaudhuri
Ph.D., M.Phil.,
Senior Professor Fisheries and Ecology Division,
Department of Zoology, University of Calcutta

2017

Scholars World

A Division of

Astral International Pvt. Ltd.

New Delhi – 110 002

© 2017 AUTHORS

Publisher's Note:

Every possible effort has been made to ensure that the information contained in this book is accurate at the time of going to press, and the publisher and author cannot accept responsibility for any errors or omissions, however caused. No responsibility for loss or damage occasioned to any person acting, or refraining from action, as a result of the material in this publication can be accepted by the editor, the publisher or the author. The Publisher is not associated with any product or vendor mentioned in the book. The contents of this work are intended to further general scientific research, understanding and discussion only. Readers should consult with a specialist where appropriate.

Every effort has been made to trace the owners of copyright material used in this book, if any. The author and the publisher will be grateful for any omission brought to their notice for acknowledgement in the future editions of the book.

All Rights reserved under International Copyright Conventions. No part of this publication may be reproduced, stored in a retrieval system, or transmitted in any form or by any means, electronic, mechanical, photocopying, recording or otherwise without the prior written consent of the publisher and the copyright owner.

Cataloging in Publication Data--DK
Courtesy: D.K. Agencies (P) Ltd. <docinfo@dkagencies.com>

Mitra, Anisa, author.
Biology and culture : potential of Chitala chitala / Anisa Mitra, Pratap Kumar Mukhopadhyay, Sumit Homechaudhuri.
pages cm
Includes bibliographical references.
ISBN 9789390384051 (Hardbound)

1. Osteoglossiformes. I. Mukhopadhyay, Pratap Kumar, author. II. Homechaudhuri, Sumit, author. III. Title.

QL638.N5867M58 2017 DDC 597.47 23

Published by : **Scholars World**
A Division of
Astral International Pvt. Ltd.
– ISO 9001:2008 Certified Company –
4736/23, Ansari Road, Darya Ganj
New Delhi-110 002
Ph. 011-4354 9197, 2327 8134
E-mail: info@astralint.com
Website: www.astralint.com

Digitally Printed at : **Sanat Printers**

DEDICATED
TO
ALL FISH LOVERS

Acknowledgements

Thankful acknowledgements are due to all the researchers who have contributed to understand the importance of *Chitala chitala* as commercially valuable species. We are also thankful to the esteemed reviewers of our publications or research paper published in different journal. The entire production team of Astral International Pvt. Ltd. deserves special appreciation for taking intent to publish our work. We want to convey special thanks to Former Professor Dr. Shanti Gopal Pal for writing kind foreword for our book.

We are also highly grateful to our family members and friends who encouraged us to disseminate information for larger public interest, persevere with and finally to publish. Last but not the least; we devote our endeavour fully to the omnipotent for the silent blessings on us all throughout.

Anisa Mitra
Pratap Kumar Mukhopadhyay
Sumit Homechaudhuri

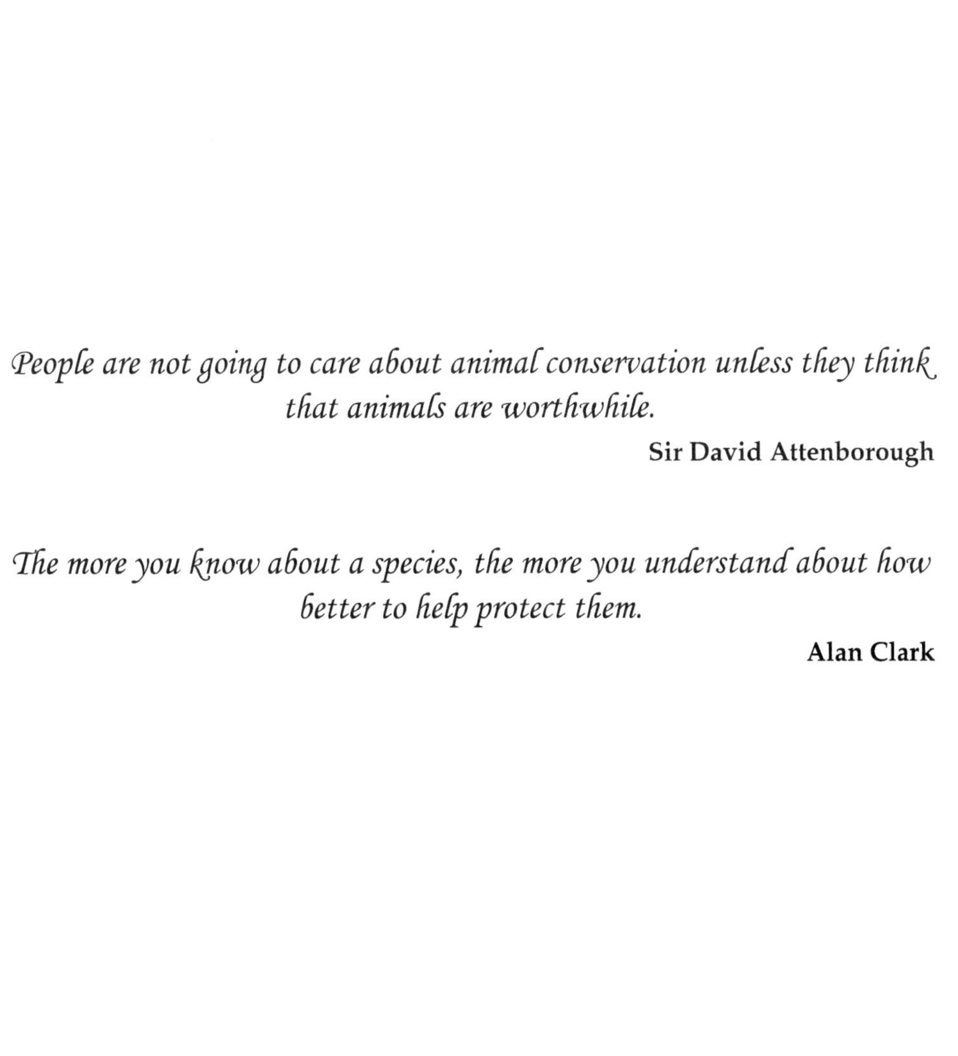

People are not going to care about animal conservation unless they think that animals are worthwhile.

Sir David Attenborough

The more you know about a species, the more you understand about how better to help protect them.

Alan Clark

Foreword

The Aquaculture industry has emerged as a dynamically developing sector in recent years, with a seven fold growth in last two and a half decades. It accounts for about half of the total fish production in our country. However, the science of fish and fisheries has an interesting traditional history and in recent years it has several significant ramifications presumably since the discovery of coelacanth, *Latimaria chalumnae* in 1938 by Dr. J. L. B. Smith. The freshwater aquaculture in our country is mainly dominated by various cyprinids comprising of major carps like Rohu, Catla, Mrigal, Calbasu and exotic carps-Silver carp, Common carp and Grass carp; catfishes like- Magur, Singhi and Pabda, the perch-Koi, medium carps like Bata and Reba. Besides, there are freshwater prawn, scampi and other small indigenous fish species (SIFS).There is still a very important fish species of the feather back group commonly known as Chital (*Notopterus chitala* or *Chitala chitala*). But somehow the species *Chitala* has not received adequate attention from aquaculturists so far despite having very high consumer preference and nutritional value. The consumer requirement is primarily met from the capture fishery from flood plains or wetlands locally known as jheels and beels. Since its regular supply cannot be assured there is a strong need to develop package of practices for its controlled culture in ponds and tanks not only as a means to diversify the cultured species and to meet its growing demand but also to ensure that such a valuable species does not become extinct. There are numerous reports available on different aspects of *Chitala chitala* but these are not compiled together. The authors have made an effort to consolidate this information i.e. its morphology, distribution and habitat, nutritional value, feeding, reproductive biology, developmental aspects and culture practices together in this work. This kind of contribution would certainly encourage the students, future researchers, as well as the aqua culturists as a standard guide to understand its biology and culture potential to uplift its present rearing protocol.

Dr. Shanti Gopal Pal
Former Professor of Department of Zoology
University of Calcutta

Prologue

Aquaculture continues to be the fastest growing animal-food-producing sector. World aquaculture is heavily dominated by Asia-Pacific region, which accounts for 89 percent of production in terms of quality and 79 percent in terms of value. The fish accounted for 17 percent of the global population's intake of animal protein and 6.7 percent of all protein consumed worldwide to outpace population growth. Globally, fish provides more than 3.1 billion people with almost 20 percent of their average per capita intake of animal protein, and 4.3 billion people with at least 15 percent of such protein. In 2015, the average annual per capita apparent fish supply in developing countries was 17.8 kg, and 10.9 kg in low-income food-deficit countries (FAO, 2016). Over the last 50 years, aquaculture became a dynamically developing and commercial business which ensures the production of desired quality/quantity "seed" at any given time by intensifying the culture regimes. It creates an income for farming families in rural or remote areas and generates jobs directly related to production or industries associated. Some 58.3 million people were engaged in the primary sector of capture fisheries and aquaculture in 2014. Of these, 37 percent were engaged full time. In 2014, 84 percent of all people employed in the fisheries and aquaculture sector were in Asia. About 18.9 million were engaged in fish farming (more than 96 percent) (FAO, 2016). Therefore, aquaculture became social and economically relevant at family, local community or country scale. But the world's capture fisheries have reached their biological limits of production and dwindling through over exploitation and habitat degradation. Hence, aquaculture not only holds promise to supply food to a growing human population but also has an important role in conservation and the recovery of threatened and endangered species by balancing the shortage of fisheries landings and by enhancing fisheries through re-stocking programmes (FAO, 2010). The key to successful management of threatened and endangered populations must be to apply appropriate conservation measures that minimize risk and maximize benefit. Therefore aquaculture is gaining importance as "last resort" in rehabilitation programmes of endangered or threatened fishes. Conservation aquaculture is an adaptive, creative approach that prioritizes preservation of wild populations along with their locally adapted gene pools and characteristics phenotypes and behaviours. It requires careful selective breeding programs to provide sufficient diversity within a fish population of interest. It is absolutely critical to understand that persistence and viability of endangered fish populations are profoundly affected by the size and structure of the population, its genetic variability and adaptive potentials. In certain situations thus aquaculture may be the only viable immediate solution for maintaining adequate effective population size (N_e) and preserving genetic diversity for long term population persistence (Anders, 1998). Previous reports showed that the recovery process for the Kootenai River population of white sturgeon (*Acipenser trnasmontanus*) in Idaho, Montana and British Columbia provides a good example of fisheries management that simultaneously incorporates conservation aquaculture and ecosystem restoration (Anders, 1998). In India fry of mahaseer (*Tor putitora*) have been successfully reared 0.2 g to 105 g in 240 days under pond environment in Uttaranchal. Sea bass, a vulnerable species has been successfully cultured in West Bengal for about 6 months by stocking hatchery produced seed. Thus timely implementation of appropriately designed conservation aquaculture programs can provide successful alternative to demographic and genetic bottlenecking, inbreeding depression and loss of unique and important locally adapted genes.

India is one of the mega biodiversity hot spots contributing about 11.72% of global fish diversity (Mittermeier and Mittermeier, 1997). India contains 662 species of animals listed as globally threatened by IUCN (2016) which is approximately 3% of the world's total number of threatened faunal species. The 662 globally threatened Indian species includes 222 species of fishes as per IUCN classification under different categories and therefore, research is being pursued recently to develop systematic conservation planning to protect freshwater biodiversity. In view of the significance and to achieve sustainable utilization of these resources, appropriate planning for the biodiversity conservation and managing strategies are of paramount importance. Hence the greatest challenge of Indian aquaculture now a days is not only to support livelihood but also to secure the fish biodiversity so that the country is able to maintain its stake on its sustainable utilization of biological resources. The introduction of local species to promote aquaculture reduces the negative impacts of exotic species and may have a simultaneous major role on the conservation strategies of endangered species (Ross et al., 2008). Candidate species for aquaculture need to meet long known biological requirements, such as good adaptability and reproduction capability under controlled conditions and rapid growth, plus socio-economic requirements, as the market acceptability and price (Hickman, 1979; Pillay, 1990). Thus the most obvious advantages when considering a new species for aquaculture are the opportunity to exploit a wider variety of ecosystems and farming areas (Ben Khemis et al., 2006) and increase efficiency by lowering production seasonality. It also widens the choice of species in the market and the attraction of new consumers (Basurco and Abellan, 1999). Altogether the culture of these new candidate species helps to maintain the price stability of existing farmed species and lower the pressure over wild stocks (Divanach and Kentouri, 2000).

India is considered as a 'carp country' due to its rich diversity of carps in its freshwater ecosystems. Carps provides the major share of Indian freshwater aquaculture, comprising around 85 percent of the total freshwater production. Carp culture in India is largely limited to six species; the three Indian 'major carps' catla, rohu and mrigal; and three exotic or Chinese carps, grass carp, silver carp and common carp (Mohanta et al., 2008). The remaining aquaculture production comes from *Pangasius* sp., other catfishes and air breathing fishes. Though recently fifteen new candidate fish species has been described in Indian aquaculture (Sreekanth et al., 2015) which can be the alternatives to the major cultured carp species, for diversification of freshwater aquaculture. To promote the sustainable use of a new candidate species with its stock enhancement and systematic conservation, a good scientific understanding on their biological attributes and culture potentialities is necessary. Here we have discussed on different biological aspects along with their potentialities to uplift the present culture protocols of the humped featherback or clown knife fish which has been prioritized as a new candidate for freshwater aquaculture system (Ponniah and Sarkar, 2000; Ayyappan et al., 2001).

Authors

Contents

Dedication *v*

Acknowledgements *vii*

Foreword *xi*

Prologue *xiii*

Abbreviations *xix*

1. ***Chitala chitala*, a New Candidate Species** **1**

2. **Geographic Distribution and Habitat** **5**
   2.1 Geographic Distribution
   2.2 Habitat

3. **General Description** **7**
   3.1 Morphological Features
   3.2 Morphometric and Meristic Characters
   3.3 Size

4. **Age and Growth Pattern** **11**
   4.1 Length-weight Relationship
   4.2 Allometric Growth Pattern During Early Ontogeny

5. **Feeding Biology** **21**
5.1 Food and Feeding Habit
5.2 Histomorphology of Digestive System During Early Ontogeny
5.3 Profile of Digestive Enzyme During Early Ontogeny
5.4 Gut Microflora and Probiotic Approach for Nutritional Supplementation

6. **Reproductive Biology** **55**
6.1 Age at Maturity
6.2 Sexual Dimorphism
6.3 Gonadal Maturity, Structure and Breeding Periodicity
6.4 Gonadosomatic Index and Fecundity
6.5 Sex Ratio
6.6 Breeding
6.7 Parental Care

7. **Developmental Biology** **63**

8. **Culture** **69**

Epilogue **73**

References **77**

Abbreviations

A.D.	anno Domini
AHA	American Heart Association
AOAC	Association of Official Analytical Chemists
APHA	American Public Health Association
A.M.	Ante Meridian
b.p.	boiling point
C	Centigrade
CAMP	Conservation Assessment and Management Plan
cm	centimeter
Conc.	Concentration
CO_2	Carbon di oxide
dH	degree of Hardness
eg	exempli gratia
et al.	et alia
etc.	et cetra
E	east
FAO	The Food and Agriculture Organization of the United Nations
FDA	Food and Drug Administration

Fig	Figure
FRP	Fibre Reinforced Plastic
g	gram
GIFT	Genetically Improved Farmed Tilapia
h	hour
ha	hector
i.e.	id est
IUCN	International Union for Conservation of Nature and natural Resources
kg	Kilogram
Kcal	Kilocalorie
Log	Logarithmic
mg	Milligram
mg L^{-1}	Milligram per Litre
ml	millilitre
mm	millimetre
MJ	MegaJoule
MS-222	ethyl 3-aminobenzoate methanesulfonate, Tricaine, 222
n	number
N	North
Ne	Effective Population size
^{o}	Degree
'	Minute
OD	Optical Density
P	Probability
P.M.	Post Meridian
PBS	Phosphate Buffer Saline
pH	Negative hydrogen concentration
ppm	Parts Per Million
r^2	R squared
SD	Standard Deviation
SEM	Standard Error of Mean
spp.	Species
SSP	Single Super Phosphate

t	ton
TL	Total length
U	Unit
V	volume
Vit	Vitamin
viz.	Videlict
vs.	Versus
%	Percentage
<	Less than
>	Greater than
µg	Microgram

1

Chitala chitala, a New Candidate Species

The humped featherback, *Chitala chitala*, one of the member of the two most primitive lineages (i.e. Osteoglossomorpha, Elopomorpha) (Near et al., 2012) of teleost, belongs to Superorder Osteoglossomorpha, Order Osteoglossiformes, Family Notopteridae. It is considered as one of the most commercially important food, sport, aquarium and highly priced cultivable fish (Sarkar et al., 2006a) due to its rarity and delicacy. Its flesh is so delicious that it is preferred by almost all fish eaters especially the belly portion. It is very rich in nutritive value (Table 1) (Gopalan et al., 2004) and contains a very good amount of fatty acids (Table 2). The profile of long chain PUFAs particularly the ratio of n3:n6 fatty acids is 1 which is ideal for human nutrition (as per the recommendation of FDA and AHA) and it may help to reduce the risk of heart disease and sudden, fatal heart attack. Hence this species can be a very good competitor of our regular foodfishes. But the smaller individuals are not valued as fresh fish and they are used as sun dried product. The soup prepared from the muscle and the oil prepared from dried fish flesh is administered to patients suffering from measles (Chonder, 1999). In addition, from the culture perspective *Chitala* has a very important role in population control of common carp, minnows, insects (Chaudhuri, 1975) and Tilapia. It can be used as a composite culture component in deep confined water (Chonder, 1999). However, over exploitation, habitat degradation, pollution and related anthropogenic pressure on their natural habitats, have considerably reduced the population of this species by 50- 60% during the last decades (Sarkar et al., 2007). Based on the CAMP report, (1998) the population of this fish has been declined more than 50% over the last

10 years and the species is categorized as endangered (EN) (Ayyappan et al., 2001; Sarkar et al., 2006 a, b, 2007, 2008, 2009a,b). However, understanding its importance and to meet high consumer demand across the country like India, Nepal, Bangladesh and Pakistan, few attempts have been made by the researchers and the farmers in recent years to revive its present population by culturing under captivity. But still, its requirement is primarily met from the capture fishery. Based on a suspected population decline approaching 30% in recent years due to over-harvesting and pollution, this species is assessed as Near Threatened, according to IUCN red list (2014). Hence, large scale farming of this species would ensure effective resource utilization, biodiversity conservation and widening consumer's choice by the diversification of freshwater Indian aquaculture (Mitra et al., 2014b).

The history of the fish is very old. This species has been found to be depicted in the ancient Indian literature, the details of it found in Sanskrit Encyclopedia-Mannasollasa (1126-1138 A.D.) and this was named as Vadisha. It was then described as a riverine, scaly, large fish feed on vegetarian diet. But later Hora (1951) identified and confirmed as Notopterus *Chitala* (Hamilton) and described it as carnivorous fish (Sadhale and Nene, 2005).It is commonly known as feather back, due to the presence of very long anal fin originating from opposite to mid pectoral and confluent with the caudal fin giving an appearance of a feather (Chonder, 1999). In Bangladesh it is known as Chital and in Pakistan it is called as Gundun. In different parts of India it is locally term as follows: Chital (Bengal), Seetul, Kandla (Assam), Mohi, Moya (Northern India), Chittala, Pulli, Pholia (Orissa), Ngapai (Manipur),Chappal mache, Cashimara, Dawana chapa, Ambutan-walah, Wallaklattah (Southern India) (Chonder, 1999).

Table 1: Nutritive value of Chital compared to some common food fishes (Gopalan et al. 2004)

Fish Varieties	Energy (kcal)	Moisture (g)	Protein (g)	Fat (g)	Mineral (g)	Carbohydrate (g)	Calcium (mg)	Phosphorus (mg)	Iron (mg)
Magur	86	78	15	1	1	4	210	290	1
Rohu	97	77	17	1	1	4	650	175	1
Puti	106	75	18	2	1	3	110	96	1
Chital	**108**	**75**	**19**	**2**	**1**	**3**	**180**	**250**	**3**
Koi	156	70	15	9	2	4	410	390	1

Table 2: Fatty Acid components of Chital compared to some common food fishes

Fatty acids	Catla	Rohu	Mrigal	Pangus	Chital
	(Jakhar et al. 2012)				(Mitra 2015b, 2016 b)
C12:0 (Dodecanoic acid)	0.24	ND	ND	0.38	**0.34**
C13:0(Tridecanoic acid)	0.15	ND	0.01	0.06	**0.04**
C14:0 (Tetradecanoic acid)	4.17	2.5	1.65	10.15	**5.1**
C15:0 (Pentadecanoic acid)	1.73	1.03	0.3	0.63	**0.37**

Fatty acids	Catla	Rohu	Mrigal	Pangus	Chital
		(Jakhar et al. 2012)			(Mitra 2015b, 2016 b)
C16:0 (Hexadecanoic acid)	38.23	32.39	32.2	34.93	**21.9**
C17:0 (Heptadecanoic acid)	2.3	1.28	0.32	0.47	**0.52**
C18:0(Octadecanoic acid)	13.22	14.15	5.09	ND	**18.2**
C19:0 (Nonadecanoic acid)	0.43	0.62	0.08	0.09	**0.07**
C20:0 (Eicosanoic acid)	0.45	0.31	0.2	0.44	**0.58**
Σsaturates	60.92	52.28	39.85	47.15	**47.12**
C16:1 n-7(Palmitoleic acid)	ND	0.67	3.51	2.27	**ND**
C16:1 n-9(cis-7 hexadecenoic acid)	ND	ND	0.44	ND	**ND**
C18:1n-9 (Oleic acid)	14.13	27.18	27.8	29.43	**21.4**
C20:1n-9(Gondoic acid)	1.13	1.7	ND	1.77	**1.34**
Σmonounsaturates	15.26	29.55	31.75	33.47	**22.74**
C18:2 n-6(Linoleic acid)	1.58	4.65	14.98	11.71	**1.64**
C 20:2 n-7 (Eicosadienoic acid)	0.31	ND	ND	0.49	**0.46**
C 18:3 n-3(α- Linolenic acid)	3.04	2.84	4.72	2.58	**3.32**
C20:3n-7 (Eicosatrienoic acid)	0.27	0.45	ND	0.79	**ND**
C20:4 n-6(Arachidonic acid0	ND	2.87	ND	0.98	**4.8**
C 20:3 n-3(Eicosatrienoic acid)	ND	ND	ND	0.41	**ND**
C 20:3 n-6(Podocarpic acid)	ND	ND	ND	ND	**3.19**
C 20:5 n-3(Eicosapentaenoic acid)	1.9	1.29	2.1	1.93	**2.7**
C 22:6 n-3(Docosahexaenoic acid)	5.4	3.74	3.76	4.48	**3.7**
Σpolyunsaturates	12.5	15.84	25.56	23.37	**19.81**
n3/n6	6.544	1.046	0.706	0.74	**1.009**

ND=Not detected

2

Geographic Distribution and Habitat

2.1 Geographic Distribution

Chitala chitala represents an appreciated fishery resources in South Asian countries like Pakistan, Bangladesh, Burma, Thailand, Malaya Archipelago, Indonesia (Menon,1974), Siam, Thailand, Philippines (DattaMunshi and Srivastava, 1988; Talwar and Jhingran 1991), Sri Lanka, Nepal and especially in India (Sarkar et

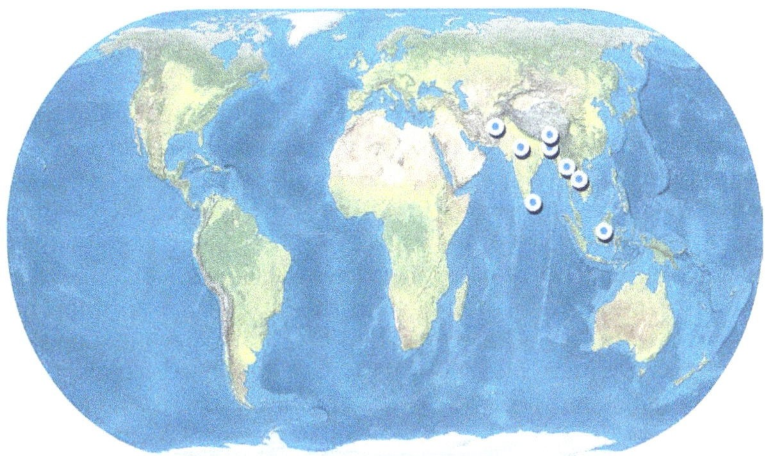

Fig 1: Map showing ( ) the worldwide distribution of *Chitala chitala*

al., 2006a, b, 2007, 2008, 2009a, b; Mitra et al., 2014a, b, 2015a, b, 2016a). However, according to Roberts (1992) *Chitala chitala* is likely restricted to the Indian subcontinent; the genus appears to be absent from Myanmar and Southeast Asia, where large sized *Notopterus notopterus, Chitala borneensis, Chitala hypselonotus, Chitala lopis* or *Chitala ornata* have been misidentified as *Chitala chitala.* In India, the distribution of the natural population has been recorded from the states of Manipur, Uttaranchal, West Bengal, Assam, Tripura, Uttar Pradesh, and Bihar (Mitra et al., 2012). According to IUCN red list (2014) the species is widespread in the western and eastern Himalayas.

In India the river Yamuna shares the large quantity and usual size in commercial catch ranges from 45 to 60 cm (Sehgal, 1973). Talwar and Jhingran (1991) observed the stray catch of chital near Kaluparaghat and in the vicinity of river Daya in the northern sector of Chilka Lake. Although the majority catch of humped featherbacks has been reported from Ganga, Yamuna, Brahmaputra, Mahanadi, Cauvery and Kalindi in India, from winter months especially from November to January (Chonder, 1999). The seed of *Chitala* can be obtained from oxbow lakes, impoundments of the rivers in the nature or from the ponds through brood stock rearing and breeding. *Chitala* is available rarely in the brackish water and in the upper tidal zone of the rivers (DattaMunshi and Srivastava, 1988; Chonder, 1999). Hossain et al. (2006) stated *Chitala* as the largest fish of Bangladesh and reported its distribution in deep and clear water of rivers, canals, beels, haors, reservoirs and ponds of Bangladesh. In India, till now riverine collection is the major sources of Chital seed. Uttar Pradesh, North Bengal, lower Assam, Tripura are the traditional hot-spot for wild stock. In West Bengal, Farraka, Dhulian in Murshidabad and Habibpur, Manickchak, Gangadharpur in Malda district and Purbasthali in Burdwan are the prime areas (Mitra et al., 2012).

2.2 Habitat

The floodplains a nd stagnant backwaters form the preferable habitats for *Chitala.* This fish species is rheophilic but it has adapted itself in artificial lakes, reservoirs and confined water. The juveniles of 110-157 mm can be found in the lake as they usually drift into the lake by the flooded rivers (Job et al., 1955; Mohanty 1975; Chonder, 1999). Talwar and Jhingran (1991) have also reported that the swamps can yield a good quantities of this species in Indian area. *Chitala* has a preference for the weedy reaches of surrounding and normally moves in shoal during day time for seeking shelter in the midst of vegetation while during night takes refuge near the bottom of the ponds. In rivers they generally dwell in well oxygenated water but it can also adapt in water with low dissolved oxygen content due to their facultative air breathing habit. This species can tolerate a high range of water temperature i.e. 6-44^{0} C and dH 5-19, pH 6-8. The young ones and adult show characteristic pattern of congregation with their heads converging together and the tails radiating outwards like petals in a flower. There is no report available on the migration of chital but it is described that the spent adults do spread away while moving upstream in the Ganga during August particularly when flood water is rising (Chonder, 1999).

3

General Description

3.1 Morphological Features

3.1.1 Morphology of Adult *Chitala*

The morphological characters of *Chitala* was well described by Day (1878). He has documented that the body of *Chitala* is superiorly coppery brown in colour with about 15 transverse silvery bars joining over the back with silvery flanks. The upper profile of the head is deeply concave with more prominent snout. The maxilla extends posteriorly behind the hind edge of orbit. The eyes with diameter 7-8 mm. in the left of head and 3/4 th of a diameter from end of snout. The preorbital is entire, with a very finely serrated lower edge preopercle. The teeth on the both jaws are villiform, and most developed on the opposite centre of the upper jaw. In vomar and palatines villiform teeth can also be found. Some teeth present on the tongue, the largest being in front. The ventral fins are minute. But the anal and caudal fins are confluent. The scales are cycloid, small, extending over the body, opercles and some of the fins. But the scales on the head are smaller than those on the body. There are about 51 serrations at the abdominal edge between the throat along the abdominal edge and the insertion of the ventral fin. The lateral line is nearly straight and present in the upper third of the body. The fins are stained with green spots, feature like black stars on the caudal region placed in a single or double row close to the anal fin and sometimes extending the whole length of its base. Talwar and Jingran (1991) reported the body of *Chitala* as oblong and strongly compressed. The head is compressed and its length about 4.2 times in standard length. The preorbital is smooth. The mouth is large with an extended maxilla beyond the hind edge of eye. The dorsal fin is inserted nearer to the base of caudal fin. The pectoral fins are moderate, extend to anal fin. The scales are small, but on opercles and on body it is of equal size. The fins are with dark blotches. Chondar (1999) has described that the adult fish has elongated and laterally compressed body with a ventral profile almost straight. The dorsal profile is strongly humped anteriorly with a marked re-entrant curve above the eyes and for this typical feature, the fish is named as humped featherback. The tail of the fish is prolonged and tapering. The snout is prominent. Mouth is terminal and the cleft is lateral, reaching below mid orbit. Two distinct spines are present, one on either side of thorax. Between throat and the insertion of the ventral fin 40-50 serrations arranged in double rows. The scales are cycloid and small. The scales on the head are slightly larger than those on the body. Dorsal fin is very small, originating slightly posterior to mid body. Pelvics are rudimentary. Anal fin is very long, originating from opposite to mid pectoral fin and confluent with the caudal fin giving an appearance of feather. Silver, coppery brown body with greyish back. Presence of 15 silvery transverse bars on either side of dorsal ridge, some conforming marks meeting on either side. Some greyish spots irregularly arranged near the end of the tail, sometimes extending the whole length of anal fin base. Dorsal fin is yellowish grey and other fins are almost white washed with silver at their basal half. Distinct markings are present along the dorsal ridge, with the characteristics black spots near the tail. But depending on the varying environmental conditions in different habitats the spots and the markings are sometimes not be seen clearly or may be absent. Sarkar et al. (2009 a) also documented that the body of *C. chitala* as oblong, strongly compressed and finely scaled. Head is compressed with the large membranous operculum flap. The mouth is wide, teeth on the jaws, palate and tongue. Eyes are large and dorso-lateral in position. Dorsal fin is small, placed in the center of back. Anal fin is razor sharp with 100-130 rays, while pelvic fin rudimentary. Vent is placed far forward. The abdominal edge is serrated with about 25-30 scutes. The number of transverse bands were recorded in the range of 16 -18. Number of black spots appeared on the anal fin as well as posterior part of the body. The characteristic spots were varied from 0 -16 on the left side and 1 - 19 on the right side.

3.1.2 Morphology of Fry and Fingerling

The morphological characters of fry and fingerling were also described elaborately by Chonder (1999). The fry of *Chitala* is measured about 15-40 mm. The body colour is silvery, with characteristics dark spots near the tail along the anal fin base. A faint, greyish, transverse bars present along the dorsal ridge. The body of the fry is laminar, strongly compressed, elongated with deeply arched dorsal contour and nearly straight ventral profile. Snout is prominent above the eye, with a deep concavity. Mouth is terminal with nearly straight large cleft reaching mid ventral orbit. Short, soft rayed dorsal fin originating slightly behind the mid body. Anal fin is long, extending posterior two third of the body confluents with the rounded caudal fin.

The size of the fingerling range from 50-100 mm. with fully developed adult characters. Dorsal profile becomes more humped than in the fry stage. The silvery ground colour of the body becoming gradually greyish to copperish along the dorsal edge. Dorsal fin at its basal half, becomes faint yellowish in advanced stage and other fins become silvery to grey. Above 80 mm. of length, a relatively prominent characteristic transverse bars present along the dorsal ridge. A varied number of dark spots become prominent near the tail along the anal base.

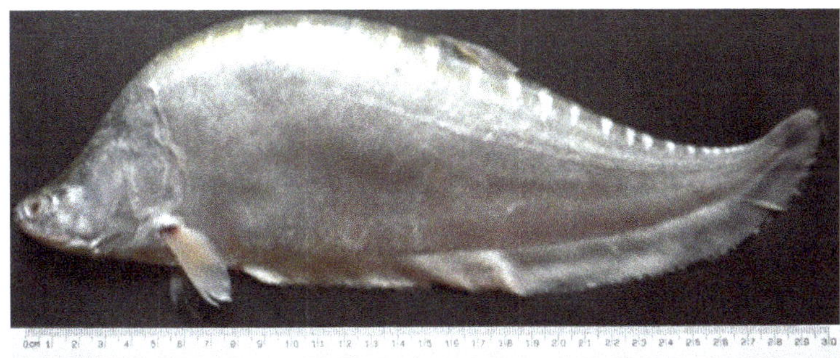

(a)

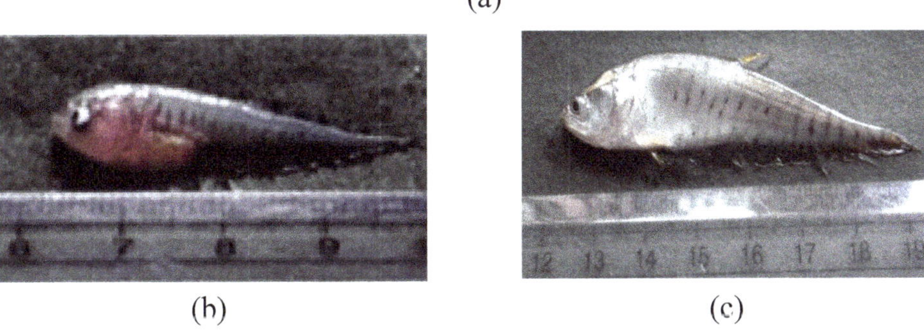

(b) (c)

Fig 2: (a) An adult humped knife (b) Fry stage of *Chitala* (c) Fingerling stage of *Chitala* (Photo credit: Anisa Mitra)

3.2 Morphometric and Meristic Characters

Many researchers earlier have described the morphometric and meristic characters of *Chitala*, all of which are summarised below: Day (1878,1889) has reported the ratio of Head length (HL) and Tail length (TL) as 4.4-5.0, the ratio of Body depth (BD) to total length (TL) is 3.5-4.0 , number of the scales on lateral line is 180 and the fin formula is D (Dorsal fin).9-10 (1-2/7-9), P (Pectoral fin).16, V (Ventral fin).6, A (Anal fin).110-125 (135), C (Caudal fin).12-14, L.1.180. Shaw and Shebbeare (1937) observed only the number of scales on lateral line was 225 and the fin formula is D.9-10, V.5-6, C.110-125 (135), C.10. Qureshi and Qureshi (1983) represented the morphometric ratio of HL and TL as 4.0-5.0, ratio of BD and TL is as 3.5-4.0.Scales on lateral line is 160-180.The fin formula is D.9-10 (1-2/7-9), P.16, V.6, A. 110-125, C.12-14. Sarkar et al. (2009a) observed the difference of morphometric and meristic characters among the intrasamples of different riverine populations (i.e. Samaspur Bird Sanctuary (SBS, 25^{0} 97'N, 81^{0}67'E), Katraniaghat Wildlife Sanctuary (KWS, 25^{0} 97'N, 25^{0} 97'E), river Ghagra (26^{0} 01'N, 83^{0}22'E), Ganga (Kanpur, 26^{0} 19'N, 78^{0} 04'E), Gomti (Lucknow, 25^{0} 90'N 82^{0} 56'E, Bhagirathi (24^{0} 05'N, 88^{0} 06'E), Farakka (24^{0} 53'N, 88^{0} 10'E), Saryu (26^{0} 12'N, 82^{0} 45'E), Satluj (31^{0} 09'N, 74^{0} 56' E), farm (24^{0} 92'N, 88^{0} 14'E) Ganga (Rajmahal, 24^{0} 80'N, 87^{0} 93'E), Yamuna (25^{0} 41'N, 81^{0} 91'E) as well as between riverine and the captive bred population of *Chitala* in India. A total 20 morphometric variables were measured namely: Total length (TL),standard length (SL),body depth (BD),Head depth (HD), Head length (HL) Length of snout (LOS): Eye diameter (ED) Predorsal length (PDL): Interorbital distance (IOD): Length of dorsal fin (LDF) Length of pectoral fin (LPF) Length of anal fin (LAF) Length of ventral fin (LPVF): Hump distance (HUMD): Width of mouth (MW) and among the meristic characters they counted number of dorsal spine, dorsal ray, pectoral spine and rays, ventral spine, anal spine, anal rays, lateral line scales, no. of black spots and dorsal bands. A significant variation was observed in case of total length, standard length, length of dorsal fin, length of pelvic fins, height of anal fin, length of anal fin, head depth, eye diameter, mouth width, hump distance and length of snout between the intrasamples of riverine populations. Between the intersamples of riverine population and captive bred population, a significant difference was also noted in case of the dorsal fin and pectoral fin length, insertion of dorsal fin and body depth. From the study they inferred that these differences among the morphometric and meristic characters are due to several factors like racial, habitat specificity and genetical. They also opined that the phenotypic plasticity in featherbacks may also be attributed to the varied environmental conditions prevailing at different locations such as temperature, food availability, prolonged swimming movement, habitat structure.

3.3 Size

Regarding the average and maximum size of Chital, several reports are available. Maximum size of 122 cm in length was reported by Day (1878). Prasad and Mookerji (1930) recorded 28.5 cm of Chital from Manchar lake of Sind. Shaw and Shebbeare (1937) obtained 91 cm long chital from North Bengal. Whereas Job et al. (1955) reported 122 cm size Chital from Hirakund reservoir of Mahanadi river of Orissa. Alikunhi (1957) also mentioned the maximum size of Chital as 122 cm. He observed that *Chitala* can achieve 300 mm. in 60-70 days in natural condition. Mishra (1959) informed that the species can attain 1219 mm. in length. Sehgal (1973) obtained 122 cm and 10.5 kg *Chitala* from The Gangetic system in Northern India. Qureshi and Qureshi (1973) stated that average length of *Chitala* is 25 cm however it may grow up to a meter. Chital can grow faster in nylon hapa fixed in pond. In pond *Chitala* can attain 360 mm. and 320 g. of weight in eight months (Singh et al., 1980). Talwar and Jhingran (1991) has reported the marketable size of the fish as 24 cm. Chonder (1999) reported that *Chitala* breeds in natural waters and the spawn takes about 30 days to attain a length of one inch and then another 70 days to grow to about twelve inches in length when it weighs about 200 g. It is possible to achieve a weight of 1 kg in eight month time. But Chital grows slowly at larval and fry stage. It grows maximum 25 mm. in 30 days under laboratory condition. Sarkar et al. (2006b) informed that a maximum of weight of 14 kg can be achieved by 122 cm of Chital.

Chitala can be captured by using seine nets from the swamp areas (Talwar and Jhingran, 1991). Different sizes of *Chitala* can be caught using grag nets, cast nets and gill nets based on different net size (Sarkar et al., 2008-2009a, b, c). *Chitala* can be attracted by using live earthworm, minnows, leeches, bee larva, maggots, night crawlers as a bait. The chopped earthworm with dried fish powder, oil cake powder, half cooked rice & some edible oil can also be used as bait for fishing. Texas rig is mostly recommended for chital fishing.

4

Age and Growth Pattern

Age and growth rates are two attributes of primary importance in accessing fish population and their response to various aspects of management measures (Bhatt et al., 2004). The purpose of age and growth pattern studies of fish is to understanding the dynamics of fish population and to determine the amount of fish productivity. Hence for successful conservation management of commercially important endangered species, precise evaluation of both age and growth pattern is very important from the culture perspective.

4.1 Length-weight Relationship

Length- weight relationship have been widely used as stock assessment model to estimate stock biomass from limited sample size, as a growth pattern indicator to compare the life histories of a species between regions and other aspects of fish population dynamics (Moutopoulos and Stergiou, 2002). Sarkar et al. (2009 c) first reported on the length-weight relationships of *Chitala* from the different wild populations of the River Ganga basin, India during the year 2000-2004. The fish samples were collected from the following river locations-Bhagirathi (24.05°N 88.06°E), Ganga (Farakka) (24°53′N 88°10′E), KWS (28°21′N 81°25′E), Ganga (Kanpur) (25°82′N 81°26′E), Saryu (26°33′N 832°37′E),Ghagra (26°75′N 81°99′E),Samaspur Bird Sanctuary (25°97′N 81°67′E),Sutluj (31°09′N 74°56′E), Malda Farm (24°92′N 88°14 ′E), Kosi (24°31′N 87°82′E). The L–W relationship of the fish was studied by linear regression of Log TW = Log a + b Log TL, where, TW was the total body weight in grams, TL total length in cm, and a and b are the parameters of the equation. A total of 221 specimens were collected ranging from 31 to 120 cm total length (TL) and 550 to 12 000 g total weight (TW). The exponential value of the length–

weight relationship 'b' of the collected samples followed the cube law, indicating an isometric pattern of fish growth in *C. chitala*. The observed variations in L–W relationships from various populations may be attributed to the different habitats on which their biology depends, including the growth phase, degree of stomach fullness, gonadal development and health condition (Sarkar et al., 2009c).

4.2 Allometric Growth Pattern During Early Ontogeny

To study the relative developmental growth pattern during early stages, morphometric ratios can also be used as a combine criteria for the qualitative evaluation of control and cultured larvae and juveniles to determine intraspecific variations (Koumoundouros et al., 1999).The allometric growth model is widely used for relative growth analysis during early larval development (Celik and Cirik, 2011) especially in the last decade with many taxa of teleost. Allometric growth model, also helps to evaluate the developmental plasticity of a particular species (Koumoundouros et al., 1999; Celik and Cirik, 2011).

In this aspect, Mitra et al. (2015 a, b) established the relationships between morphometric variables of *Chitala chitala* larvae reared at a constant temperature (30±2^0 C) from hatching to 30 days after hatching (DAH) and characterised the early developmental stages through analysis of their allometric growth. The larvae used in the study were obtained by the induced spawning of two males and one female (weight = 2.31±0.6 kg, total length = 86 ± 5.5 cm) at a local fish farm in West Bengal (India). The breeders were stocked in a pond with a surface area of 0.09 to 0.13 ha and a water depth of 150 to 170 cm. A wooden country boat was placed on the the pond bottom to provide artificial substratum. An injection of Ovaprim 1.5 ml kg^{-1} body weight was administered to the broodstock to induce spawning (Sarkar et al., 2006b), and 14 to 18 h after injection courtship and spawning occurred. The spawned adhesive eggs were removed from the pond and placed in a 100 l circular recirculating tank until hatching (168-192 h) at 27-28^0 C, pH 7.81± 0.12 and water flow 0.3 l s^{-1}. After hatching, the larvae were reared in three tanks at a density of 45,000 larvae m^{-3} (450 l). The photoperiod was maintained on a 12 L: 12 D cycle. Temperature and pH ranged from 30±2.2^0 C and 7.4 ± 0.6, respectively, throughout the experiment. Dissolved oxygen was maintained above 8.0 ± 2.5 mg L^{-1} by constant aeration. The larvae were fed ad libitum with *Artemia nauplii* three times daily. Throughout the study, daily water exchanges (10-15% of total volume) and bottom siphoning was performed to prevent the accumulation of dead *Artemia* and other organic matter. No mortality of the *C. chitala* larvae was observed during the experimental period. Sampled larvae (from hatching to 30 DAH) were subjected to overexposure in an anaesthetic solution (MS-222 at a dose of 150 mg. L^{-1}) and the morphometric characteristics were recorded. The morphometric characters and their descriptions measured during the experiment are described in Table: 3.

Table 3: Abbreviations and description of the measured morphometric characters (Mitra et al. 2014b)

Character	Abbrevation	Discription
Total Length	TL	From tip of snout to tip of the fin
Head Length	HL	From tip of snout to posterior margin of opercle
Head Depth	HD	Measure at the level of the opercula
Body Depth	BD	Posterior to the anus
Eye Diameter	ED	Measure through the centre of the eye parallel to the main axis of the body
Snout length	SNL	From the tip of the snout to the anterior margin of the eye
Tail length	TAL	Measure between the anus and the tip of the notochord
Tail Depth	TAD	Measure at the paduncle level
Pectoral fin length	PecL	Measure from the origin of the pectoral fin to its distal end
Pelvic fin length	PelL	Measure from the origin of the pelvic fin to its distal end
Trunk length	TRL	Measure from the posterior end of the operacle to the anus
Mouth Gap width	MG	Maximum opening of the mouth

All measurements were taken along lines parallel or perpendicular to the horizontal axis of the body (Gisbert, 1999). Dead and abnormal specimens (presence of malformations) were excluded from the analysis. In yolk sac larvae, the diameter (maximum and minimum) of the ellipsoidal yolk sac was also measured and the volume (mm^3) calculated using the following formula: $V = 0.1667 \quad LH^2$; where H is the minimum diameter and L is the maximum diameter of the yolk sphere (Heming and Buddington, 1988). After this, the larvae were fixed in a 4% phosphate-buffered formaldehyde (pH 7.2) solution for further detailed examinations of the form of sensory, feeding and swimming structures. Developmental stages were identified according to Kendall et al. (1984) as yolk-sac larvae (after hatching and until mouth opening), preflexion larvae (from mouth opening to the start of the flexion of the caudal tip of the notochord), flexion larvae (notochord tip reached its final position at approximately 45 degrees from the notochord axis and the principal caudal-fin rays and supporting skeletal elements are in the adult longitudinal position), and postflexion larvae (completion of notochord flexion to start of metamorphosis). The measured morphometric characters during the developmental stages of *C. chitala* are presented in Fig. 3. Allometric growth patterns during the developmental stages of *C. chitala* were modelled by a power function of TL. The patterns in allometry were described by the growth coefficient (*i.e.* power function exponent) in the equation Y = a+b, where Y is the dependent variable (measured character) and X, the independent variable (TL), a is the intercept and b, the growth coefficient (Fuiman, 1983). Isometric growth occurred when b = 1. A positive allometric growth occurred when b >1 and a negative allometric growth when b <1 (Osse and Boogaart, 2004). Linear regressions were performed on log-transformed data (TL as independent variable) and the inflection points were calculated for HL, HD, SNL, ED, MG, BD, TRL, TAL, TAD, PecL, PelL (Gisbert, 1999; Gisbert et al., 2002). Inflection points are the X values where the slope changes. The inflection points were calculated according to van Snik et al. (1997).

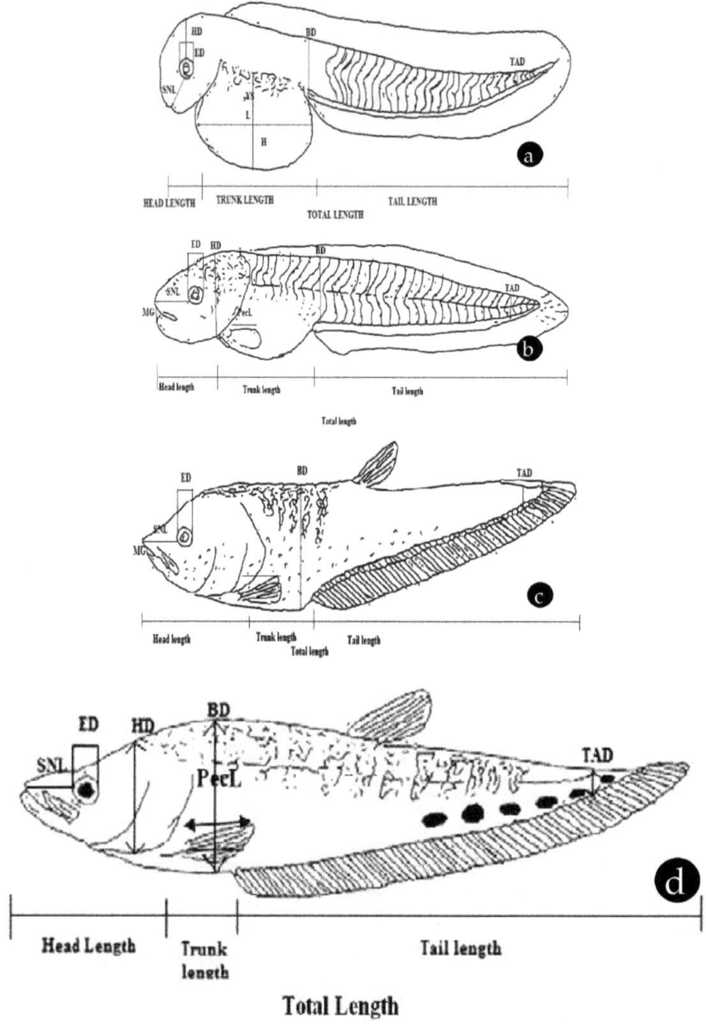

Fig 3 : Morphometric characters measured in the *C.chitala* during various developmental stages. (a) Yolk sac stage (b) Preflexion stage (c) Flexion stage (d) Post flexion stage. Abbreviations: H, minimum diameter of the yolk sphere; L, maximum diameter of the yolk sphere; YS, yolksac; SNL, snout length; ED, eye diameter; HD, head depth; BD, body depth; PecL, pectoral fin length; TAD, tail depth; TL, total length; TRL, trunk length; HL, head length; MG, mouth gape (Mitra et al., 2014b)

C. chitala followed a typical course of teleost ontogeny and the entire developmental period was differentiated into different phases according to the following morphological growth trajectories: the yolk sac phase (from 0 to 8 DAH), between hatching and yolk sac resorption and first feeding at 8 DAH; the preflexion larval phase (from 9 to 12 DAH); the flexion period (from 13 to 20 DAH); and the postflexion-juvenile phase, from notochord flexion to the completion of metamorphosis (from 21 to 30 DAH). *C. chitala* larvae measured 10.21±0.04 mm TL and grew to attain a final TL of 30.45±1.36 mm. The relation between total length and age (DAH) in *C. chitala* from hatching to metamorphosis followed a linear function (Fig.4).

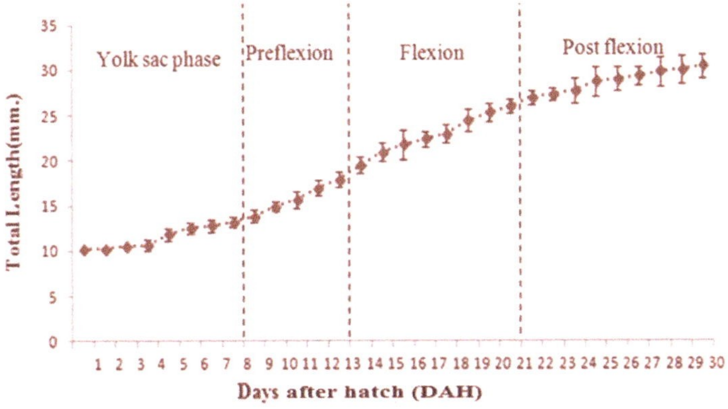

Fig 4: Growth in total length (TL) of *C. chitala* during early stages of development (from 0-30 DAH). During the period studied, growth could be defined by 8.8505+0.7793*DAH (days after hatch). Each point represents the mean ± SEM. (r^2=0·987; P<0·05; n=350) (Mitra et al., 2014 b)

During the yolk sac phase, the larvae of *C. chitala* showed an insignificant growth rate in body mass as well as length. Endogenous reserves were depleted moderately fast until mouth opening (65% of yolk and 32 % of oil globule in 8 days). Thus, the larvae showed a period of short mixed nutrition (lecithoexotrophic period) based on endogenous (lecithotrophic period) and exogenous (exotrophic period) nutrients that extended from mouth opening at 13.24 mm TL (at 8 DAH) to around 16.9 mm TL (at 12 DAH).The newly hatched larvae had the head bent ventrally with a large yolk sac (volume 3.97±0.08 mm^3)containing several oil droplets in the posterior region that migrated towards the anterior region with further development. A finfold covered the body from the dorsal area of the trunk to the ventral area of the yolk sac and was higher in the dorsal part of the trunk and narrower in the caudal region. Auditory capsules and otoliths were visible in the posterior region of the head. Scattered melanophores were observed in the dorsal part of the head and in the posterior part of the yolk sac. The eyes were darkly pigmented. The undifferentiated primordial digestive tract was observed lying dorsally to the yolk sac.

The preflexion stage began at first feeding (8 DAH) (13.24±0.73 mm TL) and ended at 12 DAH (16.9±0.92 mm TL). At the onset of this stage, the yolk sac and oil globule were almost consumed. At 8 DAH lens appeared in the eye; mouth and anus opened, and rudimentary maxilla and lower jaw bone were detected. Branchial arches were also observed from 8 DAH. The number of melanophores increased on the dorsal and lateral parts of the body. At 10 DAH (14.94±0.50 mm TL), a few small teeth were visible in the mouth. At 10 DAH with the coincidence of exogenous feeding the larvae initiated swim bladder inflation by swallowing air directly from the water surface. A convoluted digestive tract was evident from 8-10 DAH. At 12 DAH (16.9±0.84 mm TL), the dorsal, pectoral, anal and caudal finfold began to differentiate but rays were not clearly visible. The complete depletion of the yolk sac and oil globule indicated the onset of the preflexion stage (Kendall et al., 1984), during which there was a change in the growth coefficients of the head, trunk and tail segments. The head length, head depth, snout length and eye diameter growth of *C. chitala* larvae was positively allometric from hatching to 12 DAH (16.9 mm TL) and indicated continuous cerebral development, possibly allowing the improvement of the motor and sensorial ability to contribute to increase the probability of prey detection (Fuiman, 1983; Gisbert and Doroshov, 2006).

The flexion stage started from 13 DAH (17.89±0.84 mm TL) and continued until 20 DAH (25.35±1.02 mm TL). From the onset of this stage most of the larvae showed a 45 upward flexion of the notochord tip. At 15 DAH, the notochord was shortened and caudal rays filled its end. Pigmentation occurred on the bases of pelvic, anal and dorsal fins. After these stages, less drastic changes were observed.

The post flexion stage was observed from 21 DAH (25.97±1.02 mm TL) to 30 DAH (30.45±1.36 mm TL). The pectoral, anal and dorsal fins were elongated with melanophores covering most of the fins. Body pigmentation increased in the middle of the body. The first typical juvenile pigmentation pattern was observed from 25 DAH (28.78±1.6 mm TL) and continued throughout the study period. Several light and dark transverse bars were observed along the dorsal ridges. The body shape of larvae and the pigmentation pattern on the lateral side of the body resembled the 'miniature adults', but the scales were yet to form. The major morphological landmarks during larval development of *C. chitala* are summarised in Fig 5.

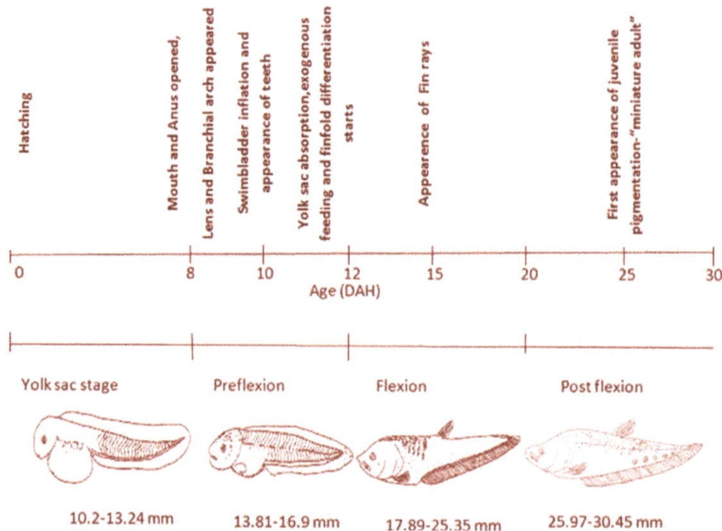

Fig 5: Major morphological landmarks in *C. chitala* early development (Mitra et al., 2014b)

Allometric growth equations between 12 measured body parts and TL during the larval developmental stage (0-30 DAH) are presented in Fig.6.

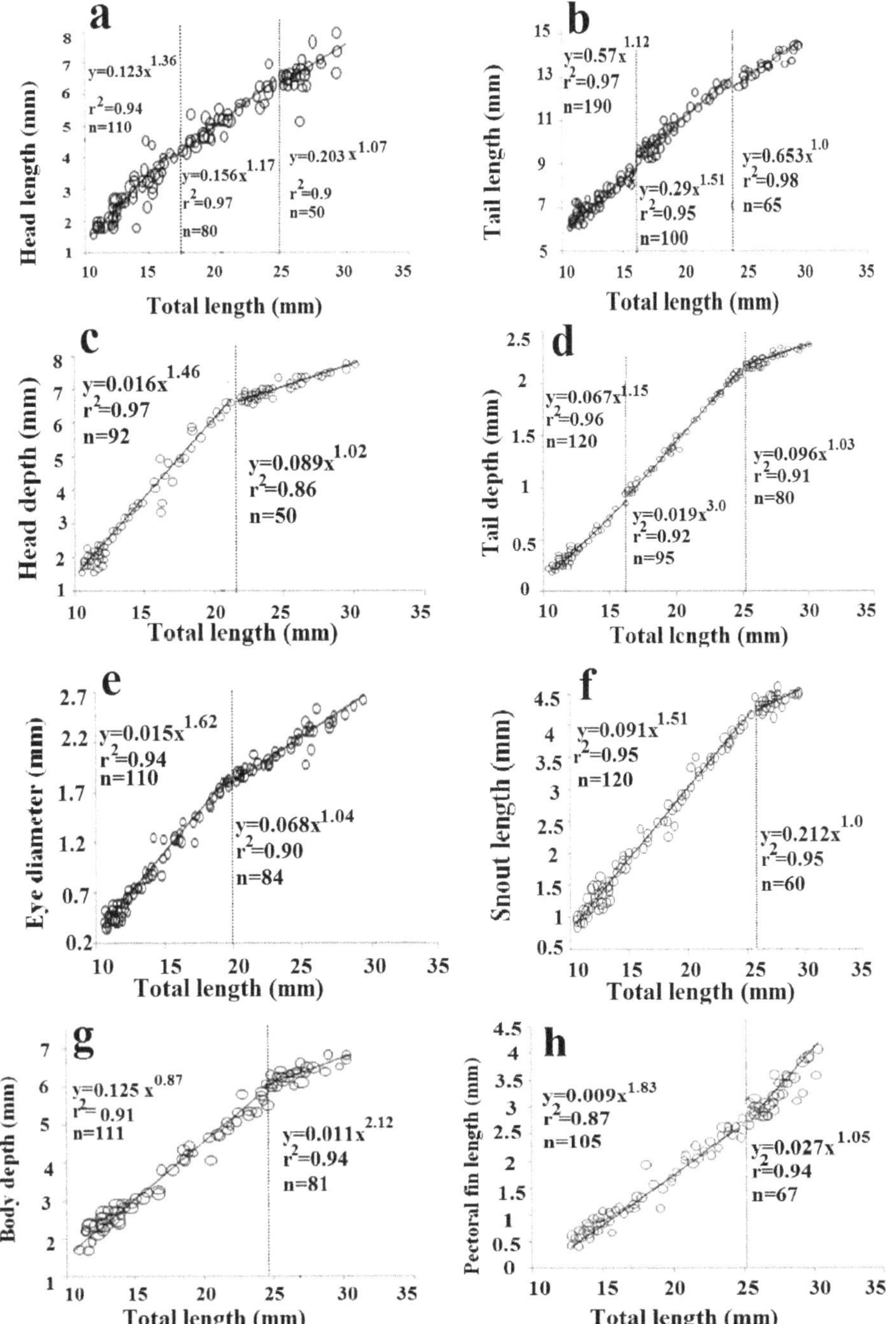

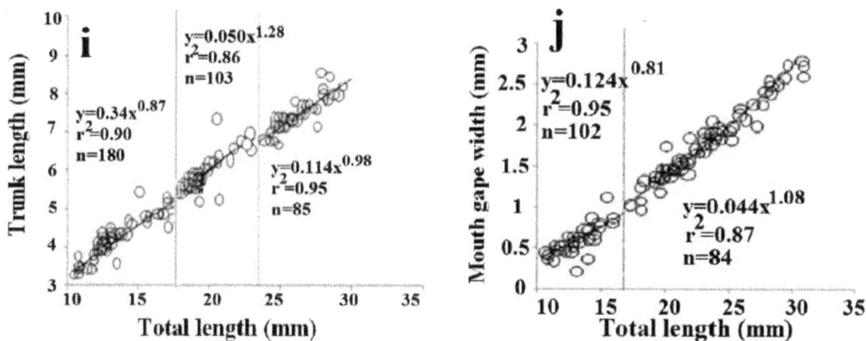

Fig 6: Allometric growth equations and relationships between different measured body proportions and total length (TL) during early stages of *C. chitala* development. The dashed line represents the inflexion point of growth. (a) Head length (first inflexion point at 16.9 mm TL, at 12 DAH; second inflexion point at 25.35 mm TL, at 20 DAH); (b)Tail length (inflexion point at 16.9 mm TL, at 12 DAH); (c) Head depth (inflexion point at 21.77 mm TL, at 16 DAH); (d) Tail depth (first inflexion point at 16.9 mm TL, at 12 DAH; second inflexion point at 25.35 mm TL, at 20 DAH) ; (e) Eye diameter (inflexion point at 19.49 mm TL, at 14 DAH); (f) Snout length (inflexion point at 25.97 mm TL, at 21 DAH) ; (g) Body depth (inflexion point at 24.47 mm TL, at 19 DAH); (h) Pectoral fin length (inflexion point at 25.35 mm TL, at 20 DAH); (i) Trunk length (first inflexion point at 17.89 mm TL at 13 DAH and 24.47 mm TL; second inflexion point at 19 DAH); (j) Mouth gape width (inflexion point at 16.9 mm TL, at 12 DAH) (Mitra et al., 2014b)

The different parameters of the allometry equations ($Y = a. TL^b$) of the morphometric characters studied are represented in the Table: 4.

Table 4: Parameters of the allometry equations ($Y = a. TL^b$) of the morphometric characters studied (Y) against TL (b allometry coefficient; a constant of the allometry equation; r^2 coefficient of determination; n number of measured individuals; ti test of isometry) (Mitra et al., 2015b)

Y	TL range (mm.)	b	a	r^2	n	ti
HL	10.21-16.9	1.36	0.123	0.94	110	+
	17.89-25.35	1.17	0.156	0.97	80	+
	25.97-30.45	1.07	0.203	0.9	50	*
TaL	10.21-16.9	1.12	0.57	0.97	190	+
	17.89-24.47	1.51	0.291	0.95	100	+
	25.35-30.45	1	0.653	0.98	65	*
TAD	10.21-16.9	1.15	0.067	0.96	120	+
	17.89-25.35	3	0.019	0.92	95	+
	25.97-30.45	1.03	0.096	0.91	80	*
HD	10.21-21.77	1.46	0.016	0.97	92	+
	22.41-30.45	1.02	0.089	0.86	50	*
ED	10.21-19.49	1.62	0.015	0.94	110	+

Y	TL range (mm.)	b	a	r^2	n	ti
	20.94-30.45	1.04	0.068	0.9	84	*
SNL	10.21-25.35	1.51	0.091	0.95	120	+
	25.97-30.45	1	0.212	0.95	60	*
PecL	10.21-25.35	1.83	0.009	0.87	105	+
	25.97-30.45	1.05	0.027	0.94	67	*
BD	10.21-24.47	0.87	0.128	0.91	111	-
	25.35-30.45	2.12	0.011	0.94	81	+
TRL	10.21-17.89	0.87	0.34	0.9	180	-
	19.49-24.47	1.28	0.05	0.86	103	+
	25.35-30.45	0.98	0.114	0.95	85	*
MG	10.21-16.9	0.81	0.124	0.93	102	-
	17.89-30.45	1.08	0.044	0.87	84	*

+, positive allometry; -, negative allometry; *, isometry

The positive allometric growth of eyes is considered an indicator of development and differentiation of neural and sensorial structures; which would allow the larvae to react to light stimuli, detect zooplankton prey and potential predators in the water column (Gisbert et al., 2002; Gisbert and Doroshov, 2006). An increment in the growth coefficient of the head segment during preflexion stage can also be correlated with the development of the respiratory (branchial filaments) system in *C. chitala* at 8 DAH. During most of the fish larval stage, cutaneous respiration is known to be more important than gill respiration (de Silva, 1974).The development of gill arches and filaments allow switching from temporary (cutaneous) to branchial respiration, resulting in a better oxygen supply and an increase in swimming activity (Gisbert et al. 2002). The swim bladder inflation also occured around 10 DAH in *C. chitala* larvae, which seems to have a crucial function for controlling buoyancy and swimming activity and also for capturing prey more efficiently and avoid predators during larval development (Russo et al., 2007). The first growth period ended at the beginning of differentiation of primordial finfold shortly before the notochordal flexion stage, which indicated that the larvae began to modify their swimming mode (Bone et al., 1995). A significant increase in tail depth occurred concomitantly with the development of unpaired fins and fin rays in *C. chitala* during this stage. This increment in tail depth, may correspond to the development and differentiation of the hypural plate (Ortiz-Galindo et al., 2000) and the caudal peduncle (Johnston and Hall, 2004). The comparative enhanced development of tail depth in *C. chitala* also implicitly indicates the enlargement of propulsive area and the increase of propulsive power (Fuiman, 1983) with the development of two gear red and white muscle system from a single dominant fibre type. This results in a considerable change in body shape from an elongated preflexion larva to a more robust flexion specimen (Blaxter, 1988). The morphological differentiation of the caudal fin and development of lepidotrichia in *C. chitala,* particularly in this growth period closely paralleled the progressive change in the larval swimming mode from the anguilliform motion to the subcarangiform swimming (Bone et al., 1995; Osse and Boogaart, 1999). These changes and shift in swimming strategy helps the larvae to enhance swimming efficiency (Muller and van Leeuwen, 2006) and thus reduces the transport costs (Blaxter, 1988; van Snik et al., 1997; Muller and van Leeuwen, 2006) for both food capture and predator avoidance (Williams et al.,1996). Kolmann and Huber (2009) stated that a positive allometry in feeding performance assists predators in overcoming the functional constraints imposed by their prey, and may confer a competitive advantage over isometric ontogenetic trajectories, facilitating access to exclusive trophic resources earlier in life. Thus, the positive allometric growth of the mouth gape in *C. chitala* during this stage helps the larvae to improve their prey capture ability (Osse and Boogaart, 2004).*Chitala chitala* being a predator fish (Rahman, 1989) shows a slow overall movements followed by a rapid strike, like other predatory fishes (Kammerer et al., 2005). According to Walker (2004), these actions are commonly associated with the involvement of both caudal fin movements to generate an impulse and pectoral fins for maneuvering. The study revealed that pectoral fins showed a positive growth allometry during the flexion period in *C. chitala* which were attributed to their crucial function in swimming and maneuvering for feeding. According to Betti et al. (2009) the pectoral fins are the first to appear but the last to obtain a full complement of rays, which aid in locomotion and prey capture in larval teleosts (Osse and Boogaart, 2004). Being a primitive basal teleost (Mandal et al., 2012) the development of pelvic fin was rudimentary in *C. chitala* during the studied period. Murata et al. (2010), also proposed that primitive teleosteans often have extremely limited pelvic fin function compared with more derived fish, where the pelvic fins have a trimming function that reduces pitching and upward body displacement during braking. The extended low initial growth in *C. chitala* larvae during the first developmental phase can be described by the fact that the available energy was allocated to enhance the vision, improve feeding capabilities and swimming ability. These in turn allow them to attain a better nutritional status (i.e. changes in the prey type selection, shifting from small to larger more nutritious prey) (Morote et al., 2008) to increase the body mass in the latter developmental stages (Parra and Yufera, 2001). A significant morphogenesis and occurrence of growth in the trunk region of *C. chitala* larvae may also be due to the differentiation and growth of myotomes and the development of digestive organs (Gisbert and Doroshov, 2006). *C. chitala* showed a decrease in growth rate approaching isometry during the latter stages of growth, which is a typical feature of a juvenile fish growth profile, and defined the end of metamorphosing period (Fuiman, 1983). This change to isometry has been considered as a natural transition in growth priorities since primary functions have been fulfilled during the early developmental stages (Osse and Boogaart, 2004).

5

Feeding Biology

The feeding biology of fish is one of the important aspects of the fish biology. The knowledge of feeding biology helps to produce optimum yield by utilizing all the available potential food of the water bodies properly without any competition. Feeding habits of fish help to determine the ecological condition of fish, niche in the ecosystem and preferred food items in order to obtain a correct picture of nutrition and feeding adaptations (Polling, 1993).

5.1 Food and Feeding Habit

A sufficient number of reports are available regarding the food and feeding habit of *C. chitala* from India. According to Alikunhi, (1957) *Chitala* changes their feeding habit completely at early stage and starts feeding voraciously on carp fry and aquatic insects. Sharma and Chandy (1961) studied on the alimentary canal and feeding habit of *C. chitala* and described it as a carnivore. They reported that the specialized type of dentition and highly developed teeth like gill rakers in *Chitala* are precisely fit to give a faster and good grip to prey. The opening of the mouth is large to capture larger prey. The buccal cavity lining consisted of stratified epithelium, sub epithelial connective tissue and muscularies. The high rate of mucus lubcrication from the abundant mucus cells of buccal cavity, pharynx and esophagus helped the fish to soften the prey before entering into the stomach. The stomach was highly developed and pear shaped, which enabled the fish to digest small fish, insects quickly. The stomach possessed a single layer of columnar epithelium and a broader muscular coat. Presence of a major part of cardiac stomach suggested a greater need for digestive activity of the animal food by the species. The stomach was without any blind sac and leads to a short, single coiled intestine. The pyloric appendages were very long and two in number with internally present transverse rugae which indicates the species as carnivore. Single recctal caecum was present. In the columnar epithelium of the intestine a number of mucus cells were present while the rectum is devoid of such cells. Das and Moitra (1963) also reported that the occurrence of animal material was 80% and plant matter was up to 15 % in the diet of *C. chitala* indicating its preference for animal prey. Parmeshwarn and Sinha (1966) showed that the occurrence of different types of food items in the gut of *Chitala* depending on their availability. Ghosh (1985) observed that in adult *Chitala* the anterior intestine has higher amount of lipase but the activity of protease, amylase and invertase was weak. Whereas the posterior intestine shows only lower amount of protease and lipase. The stomach had higher quantity of protease and weak lipase activity. The pyloric caeca secreted only a moderate amount of protease, amylase and lipase with a very low invertase. But the rectal caecum was lacking of any enzyme. He also reported that the acidic pH of stomach ranges 3.0-5.8, anterior intestine 6.4-7.5, posterior intestine 6.8-7.5 and pyloric caeca 7, which helps the fish to digest animal food. Sarkar et al. (2006a) described the dietary preference and feeding behaviour of early life stages (fifteen days old post hatchlings) of *Chitala* under laboratory condition (in a 30 L recirculatory tank) in terms of their survival rates, specific growth rate (SGR), percentage of weight gain. They used eight different diets i.e. live feed (tubifex worms, chironomous larvae, zooplanktons,), dry feed (dry tubifex, spirulina, daphnia) and other non-conventional feed (fish eggs and boiled egg-yolk) during the experiment. The fishes accepted all types of diets, but the higher feed intake was observed for live tubifex, fish eggs, chironomous larvae and plankton. The highest specific growth rate (SGR) was observed in case of live tubifex worms. Sarkar et al. (2006a) opined that the reasons may be due to the wriggling movements of large and nutritionally rich prey organisms like tubifex and chiromonous larvae which minimize the temporal and energy cost of feeding and maximize growth in larvae. But there was no difference of mean survival in case of live feed (tubifex, chironomous larvae, zooplanktons) and artificial diet (dry tubifex, spirulina, daphnia). They also noted that the larvae continuously moved upward and downward directions showing typical preying habit of swallowing feed. No cannibalism was noticed but the fishes became sluggish after feed intake. Larvae were negatively phototactic and took shelter in the artificial hidings. In another experiment performed by Sarkar et al. (2007) where they first accounted the comparison on the feeding, growth (% weight gain and SGR) and survival rates of early life stages of *Chitala* between live (tubifex worms, chiromnous larvae, zooplanktons), dry feed (dry tubifex, spirulina, daphnia) and nonconventional feed (fish eggs and boiled egg yolk) in two different rearing systems i.e. in hapa (7x 3x 1.5 ft) and in 200-lit FRP tank. The early life stages of *C. chitala* showed good adaptability, growth and survival in both the rearing systems. In the hapa, the larvae fed on live tubifex a higher value of SGR and weight gain per cent in comparison with fish eggs and IMC spawn. But in FRP tanks the ELS fed on chironomous larvae, mosquito larvae and live tubifex showed a rapid growth rate compared to zooplanktons. A higher value of SGR was also observed in larvae fed on chironomous larvae, while the lowest value was observed for zooplanktons. During the experiment it was noted that the larvae were easily adaptive to captive conditions and immediately accepted the diets. They showed a continuously moving behaviour in circle in the middle of the water column. It was also revealed by experiment that the eggs of *Mystus vittatus* served as excellent diet for rearing post-hatchlings of *C. chitala* which might be due to its high protein contents. The experiments showed that early life stages of *C. chitala* could be reared successfully in hapas and fibreglass reinforcement tanks to attain better growth and survivability. But there are many factors such as the production system, type and size of rearing tanks, the size of fish and the quality and quantity of food and stocking density and this should be taken into account while rearing the fishes in controlled environment (Mgaya and Mercer, 1995). Sarkar et al. (2007) also opined that the ability to use non-live diets by *C. chitala* observed during the experiment has an immense commercial importance which can increase the economic viability by drastically reducing the use of live prey and labour costs.

Sarkar et al. (2009b) described the diet and feeding habit of *Chitala* by analysing the stomach content and measuring Gastrosomatic index collected from different wild population of Samaspur bird sanctuary (25^0 97′N 81^0 67′E), Kanpur (25^0 82′N 81^0 26′E), Saryu (26^0 33′N 832^0 37′E), Gomti (25^0 90′N 82^0 56′E), Katerniaghat wildlife sanctuary (28^0 21′N 81^0 25′E), Ghagra (26^0 75′N 81^0 99′E), Bhagirathi (24^0 05′N 88^0 06′E), Ganga (Farakka) (24^0 53′N 88^0 10′E), Sutluj river (31^0 09′N 74^0 56′E) during premonsoon (February- May), Monsoon (June – September) and post monsoon (October- January) during the year 2000-2005. They showed that the alimentary canal of *C. chitala* was short, muscular, bag shaped, less coiled and the stomach content with high incidences of crustaceans, molluscs and fishes indicating this species as a carnivorous predator. The study also reported that *Chitala* is capable of widening the food spectrum depending on the availability in the aquatic ecosystem. The samples collected from the river Gomti and Saryu, with a variation of habitat structure had the maximum value for feeding index. Whereas due to the fast flowing nature of the river Gerua and with less abundance of natural food organism the lowest feeding index was recorded from there. The maximum value for gastrosomatic index from February to May (premonsoon) indicated that the increment in feeding activity and the lowest value during June to September (monsoon) coincided with the spawning season. Kundu and Homechaudhuri (2014) described that the prey preference of *Chitala* increased with progression of developmental stages and the juveniles exhibited nocturnal, voracious predatory activity. The study also reported that during day time, the specific clumped distribution pattern of juvenile fishes in open and closed environment acts as a survival strategy to minimize the chances of predation and mortality in juveniles, irrespective of food availability.

5.2 Histomorphology of Digestive System During Early Ontogeny

The comprehensive knowledge on the structural and functional changes during the development of the larval digestive tracts, is of principal importance in optimizing rearing strategies for a particular species. This knowledge can help to identify limiting factors during larval rearing, reduce bottlenecks in the weaning process, and optimize feeding technology practices (Zambonino-Infante et al., 2008).

The establishment of appropriate moment for beginning exogenous feeding during the development is also a key factor from practical and economical perspective. Mitra et al. (2015a) focused on the digestive system development with descriptions of morphological and histological changes during the ontogeny of featherback from hatching to 30 days after hatch (DAH).The larvae used in the study were obtained by following the similar protocol of fertilization and rearing described by Mitra et al. (2014b). The sampled larvae (0-30 DAH) were anesthetized with MS-222 (35 mg l^{-1}), rinsed in distilled water, and fixed in neutral-buffered formaldehyde. A subsample of 15 larvae was individually measured to the nearest 0.01 mm using a light microscope, and after this the larvae were weighed to the nearest 0.001 g (wet weight) with an analytical microbalance and mouth width (gape) was measured under stereomicroscope with ocular micrometer. For histomorphological analysis, the remaining 35 larvae were fixed in Bouin's solution, dehydrated through a series of alcohol concentrations, cleared in xylene, and submerged in paraffin wax, and cut into 5 µm sections. Serial, sagittal, and transverse sections were stained following Harris's Haematoxylin and Eosin (HE) procedure for general histomorphological observations (Pearse, 1985) and viewed under a light microscope to describe the development of the digestive tract.

The yolk sac stage was coincided with the lecithotrophic (endotrophic) period i.e. from 0 to 8 DAH; the lecithoexotrophic (endoexotrophic) period was observed from 8 to 12 DAH i.e. from the beginning of preflexion stage and ended with the commencement of flexion stage; the exclusive exotrophic period, was observed from 12 to 30 DAH i.e. from the start of flexion stage and continued with the postflexion stage. The survival percentage of the larvae at the end of the experiment was 80 %.

5.2.1 Growth and Mouth Width During Early Developmental Period

The relationship between total length, body weight (wet) and mouth width (MW) per larvae related to days after hatching during the studied developmental period of *C. chitala* is presented in Fig. 7. The relationship between mouth width (MW) and total length (TL) was linear and body weight (BW) is proportional with total length (TL):

$$MW = 0.1671\ TL - 0.2345 (R^2 = 0.98)$$

$$TL = 1.465\ BW\ 0.4783\ (R^2 = 0.9855)$$

In *C. chitala* larvae a slow growth was observed during 1–18 DAH, but dramatically increased from 19 to 30 DAH during the post flexion stage. The linear relationship between mouth width and body length in the present study suggests that *C. chitala* can be a good candidate for early weaning with compound diet. The mouth width (gape) increased during development and reached to 2 mm in 30 DAH larvae and the available technology allows adapting the microparticle size to the mouth width, starting from 50 µm (Cahu and Zambonino-Infante, 2001). Beside convenient technology, knowledge on nutritional requirements in *C. chitala* is needed to formulate suitable compound for this species. Thus further study on the digestive specificities of early stages will ensure to adapt the nutrient nature and its form of supply in microdiet and the microparticulated diets might be formulated taking into account the capacity of larvae to digest this diet.

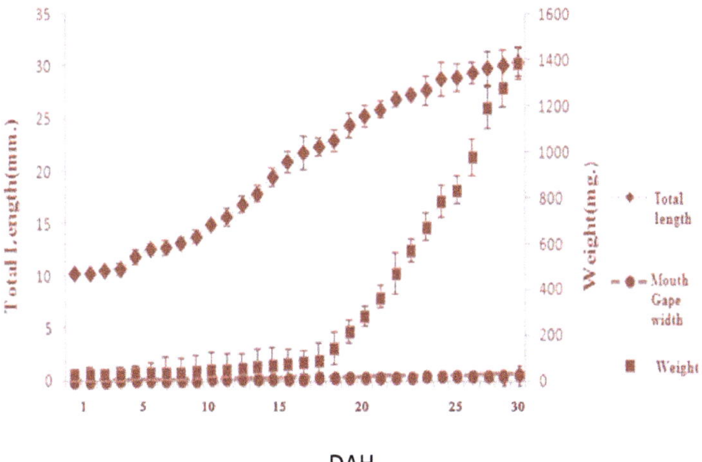

Fig 7: Growth in total length and weight of *C. chitala* larvae from hatching until 30 DAH (days after hatching). Each point represents the mean of fifteen measures ± SEM. (Mitra et al. 2015a, 2016a)

5.2.2 Gut Histomorphology

The intense larval histomorphogenesis was observed mostly during the first two stages, while during stage 3 only the size and the complexity of pre-existing organs increased. The histomorphological changes during the larval development are described below.

5.2.2a Endogenous Reserve

At hatching, featherback larvae had a large yolk sac lined by a vitelline envelope and attached beneath the anterior half of the body. The undifferentiated primordial

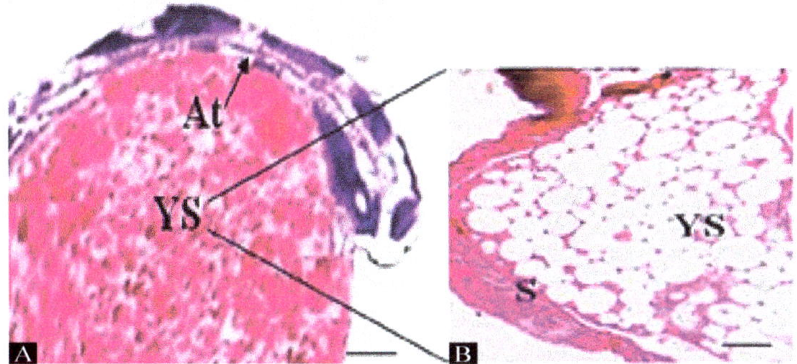

Fig 8: Histological sections of Yolksac larvae of *C. chitala*. (A) Sagittal section of *C. chitala* larva at 1 DAH. The alimentary tract is undifferentiated and mouth is closed. At - alimentary tract; Ys - yolksac. Scale bar 200 µm. (B) Histological sections of the yolk sac in *C. chitala* larvae. Yolk globules surrounded by the yolk sac syncytium (S). Scale bar 50 µm. (Mitra et al., 2015a)

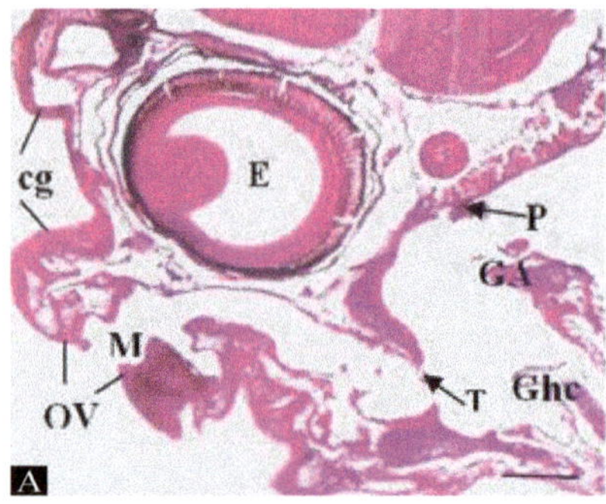

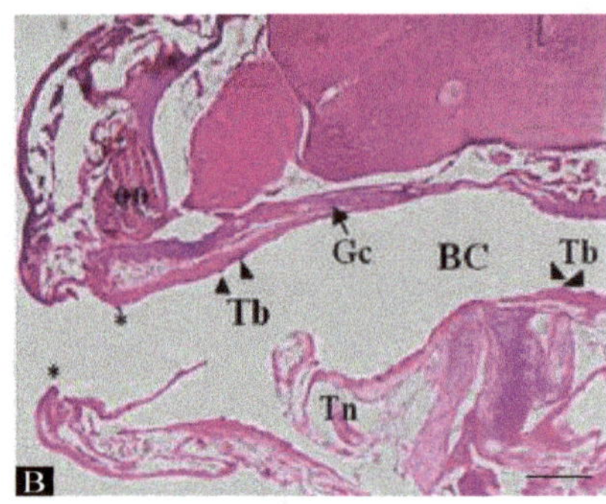

Fig 9: Histological sections of buccopharynx at different developmental stages of *C. chitala.* (A) Sagittal section of *C. chitala* larvae at 8 DAH. The mouth is open and oral valves are present. (B) Sagittal section of the buccopharyngeal cavity during the development of *C. chitala* larvae at 10 DAH. The buccopharyngeal cavity showing the presence presence of jaw teeth (*), taste buds, olfactory organ and goblet cells in buccopharynx. Abbreviations: cg—cement glands; E—eye; ga—gill arches; M- mouth; OV-oral valves; P- pseudobranch; T- teeth; Tn-tongue; Ghc- glossohyal cartilage; BC- buccopharyngeal cavity; Tb- Taste bud; GA- gill arch; Gc- goblet cell ; OO - olfactory organ. Scale bar 200 µm. (Mitra et al., 2015a)

digestive tract lay dorsally to the yolk-sac and did not communicate with the exterior because the mouth and anus were not yet opened (Fig. 8a). The yolk sac contained several acidophilic yolk globules surrounded by a syncytial epithelium (Fig. 8b). There were many blood vessels around the yolk. The mouth and anus opened on 8 DAH synchronously with the partial absorption of the maternal reserves. The yolk sac volume depleted gradually until complete absorption by the end of 12 DAH.

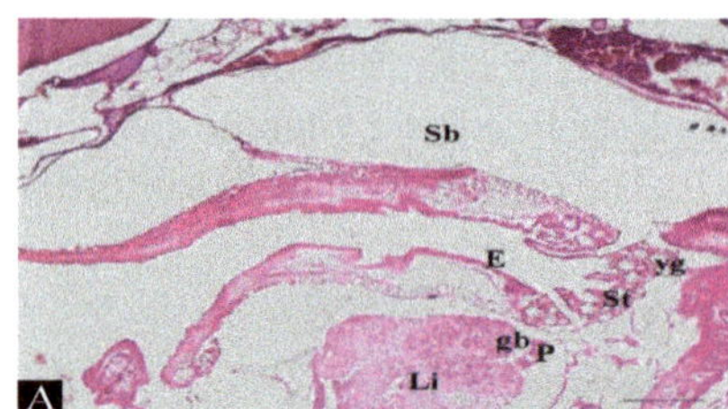

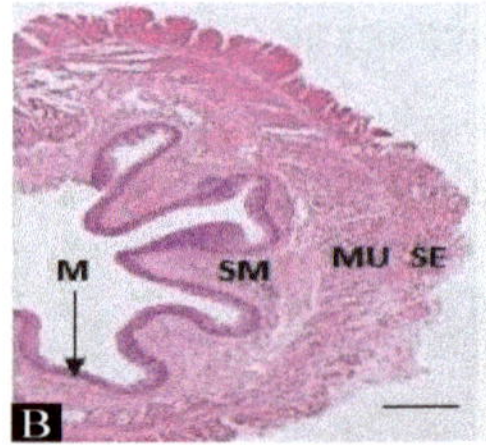

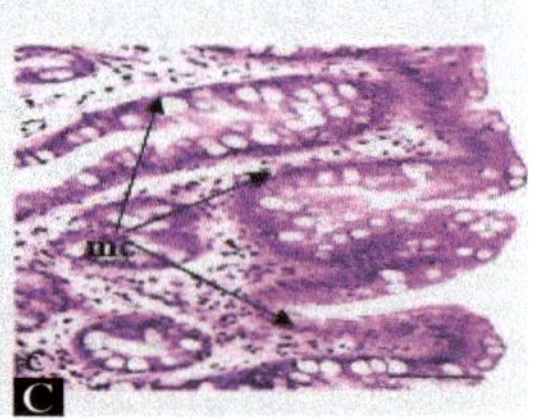

Fig 10: Histological section of esophagus at different developmental stages of *C. chitala.* (A) General view of the transition between the esophagus and gastric stomach showing longitudinal folds surrounded by the yolk globule (yg) at 10 DAH. Arrow indicates the junction between swimbladder and esophagus. Scale bar 200 µm. (B) Details of esophagus at 12 DAH larvae, showing four layers, Serosa (SE); Muscularis (MU); Submucosa (SM) and Mucuosa (M). Scale bar 100 µm. (C) Abundance of mucus cells in posterior esophagus 20 DAH.Abbreviations: E - Esophagus; Li - liver; I-intestine; St- stomach; P- pancreas; gb – gallbladder; Sb - swim bladder; mc- mucus cells. Scale bar 50 µm. (Mitra et al., 2015a)

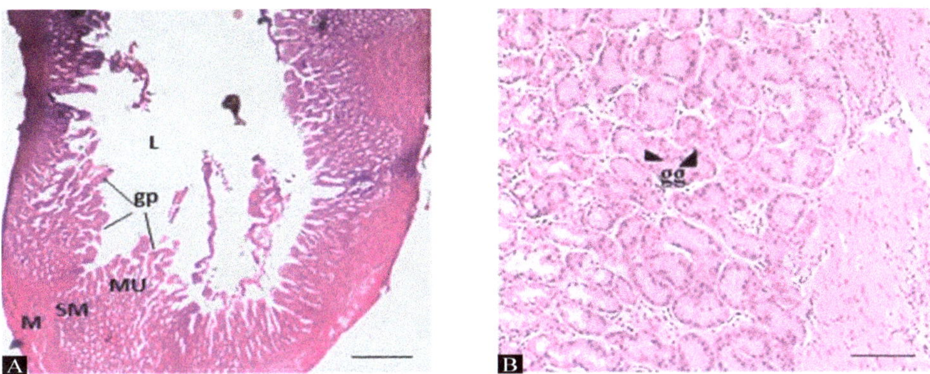

Fig 11: Histological section of stomach at different developmental stages of *C. chitala*. (A) General histological view of stomach wall at 15 DAH *C.chitala* larvae showing L-lumen; gp -gastric pits; mucosa; SM- submucosa and M-muscularis. Scale bar 200 µm. B Histological section of the glandular fundic stomach of 25 DAH *C.chitala* larvae, showing numerous gastric glands (gg). Scale bar 50 µm. (Mitra et al., 2015a)

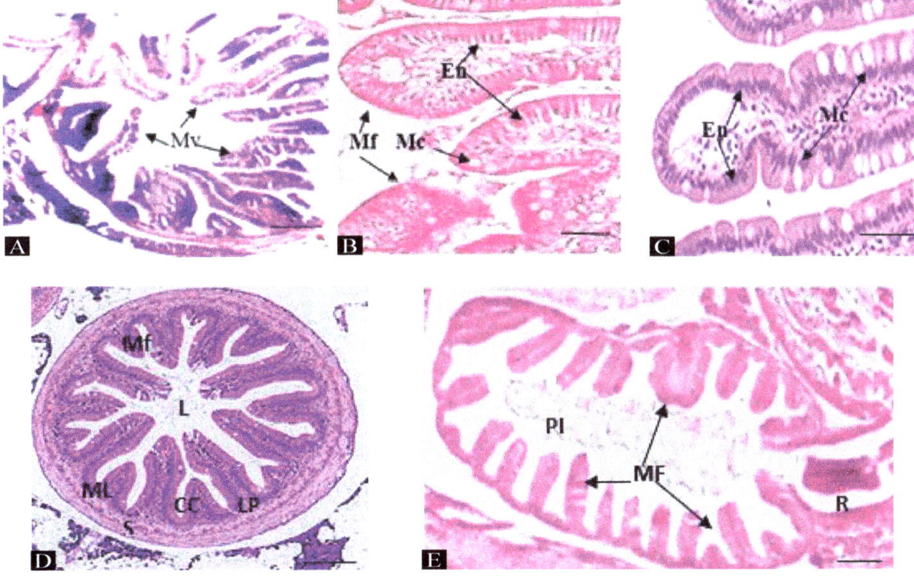

Fig 12: Histological sections of intestine at different development stages of *C.chitala*. (A) Histological section of intestine of *C. chitala* larvae on 6 DAH showing the microvilli. Scale bar 200 µm. (B) and (C) Details of the mucosal folds (Mf) with enterocytes (En) and mucous cell (Mc) on 12 and 20 DAH .Note the increase in mucus cells on 20 DAH compare with 12 DAH. Scale bar 100 µm,50 µm. D Transverse sections of pyloric caeca in 12 DAH *C. chitala* larva showing lumen (L); mucosal fold (MF); and columnar cell (CC); lamina propria (LP); muscularis layer (ML); serosa (S). Scale bar 200 µm. E Detail of posterior intestine and rectum in 12 DAH *C.chitala* larvae showing mucosal folds (MF). Scale bar 200 µm. (Mitra et al., 2015a)

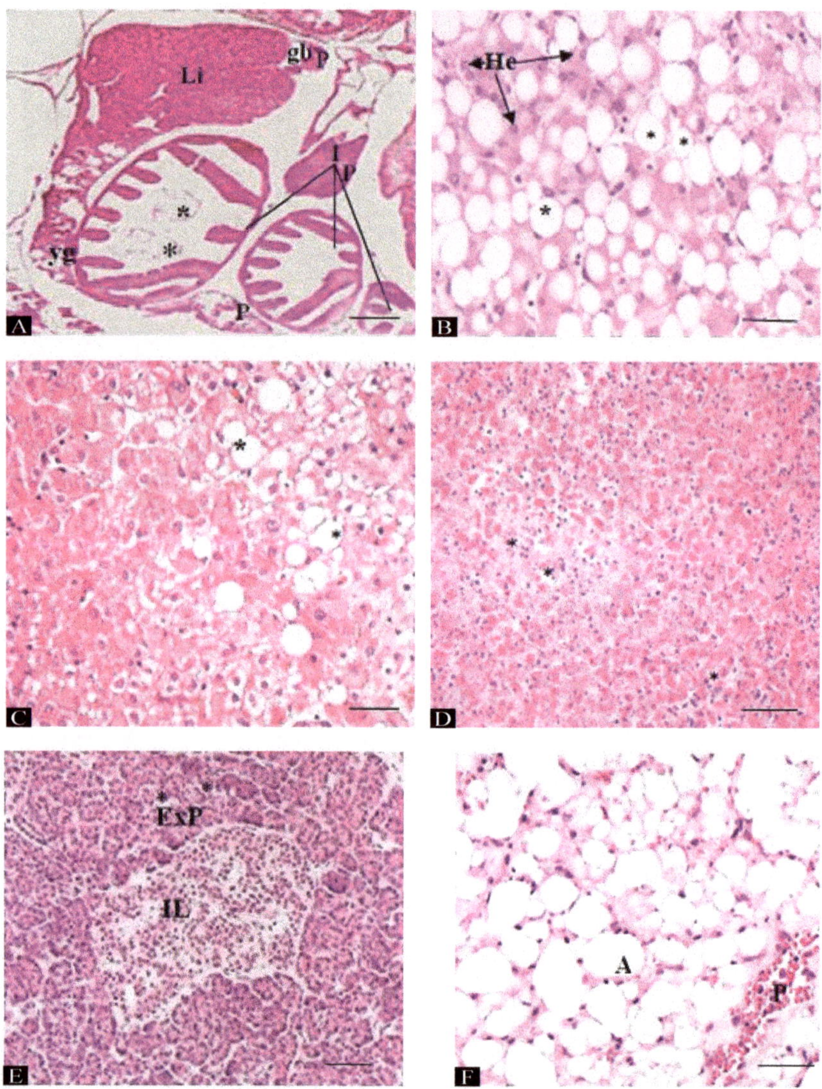

Fig 13: Histological sections of the liver and pancreatic tissue of *C. chitala* at different stages of development. (A) Sagittal section of larvae at 8 DAH, showing developing liver and pancreas with gall bladder, ingested food material in gut (*). Scale bar 200 µm ((B) Liver in a 6 DAH larvae filled with large yolk globules (*) within pigmented hepatocytes; (C) Reduction in fat deposition within hepatocytes at 10 DAH; note the reduction of fat deposits between both ages with regard to 6 DAH; (D) Regular arrangements of hepatocyte on 12 DAH showing eosinophilic granule indicating glycogen deposits (*). Scale bar 50 µm, (E) Pancreatic tissue section at 8 DAH larvae, showing exocrine pancreas (ExP) and islets of Langerhans (IL). Note polyhedral basophilic pancreocytes arranged in acini containing zymogen granules (*). Scale bar 50 µm (F) Adipose tissue (mesenteric fat deposits) surrounding the pancreas at 25 DAH larvae. Scale bar 100 µm. Abbreviations: A - adipose tissue; IL - endocrine pancreas (islet of Langerhans); P - pancreatic tissue; Li - liver; I – intestine; gb – gallbladder; yb – yolkglobule; mc -mucus cells. (Mitra et al., 2015a)

5.2.2b Mouth, Oral Cavity, and Pharynx

At hatching (TL= 10.21 ± 0.03 mm), the mouth was closed by a covering membrane, and the jaw cartilage was not formed. The mouth opened on 8 DAH, and two oral valves were evident in the larvae (Fig. 9A). Cement glands were obvious and located anteriodorsally to the head. At this age, the medial glossohyal cartilage, which would constitute the future tongue, was also evident on the ventral oral valve. The buccopharyngeal epithelium was composed of a single layer of squamous cells with scattered round mucous cells, mostly found in both oral valves. The buccal cavity and pharynx were separated by the primordial gill arches, which is supported by cartilage and was also appreciated at this age. The first goblet-shaped mucous cells appeared at 10 DAH (TL = 14.94 ± 0.5 mm) dispersed within the stratified squamous epithelium. The oral cavity was short, and canine-like teeth were detected on the oral valves protruding into the buccopharyngeal lumen. A few scattered taste buds were also visible in the buccal cavity epithelium at this age (Fig. 9B). With the larval development the size of the oral cavity increased, the number of teeth, mucous cells, and taste buds increased substantially. Four gill arches with pseudobranchia were visible at 8 DAH proliferating toward the pharyngeal cavity. The gills were completely formed at 12 DAH with filaments of increased length. The primordial lamellae increased in number and size throughout larval development. No noticeable histological changes were observed after 12 DAH until the end of the study at 30 DAH (TL = 30.45 ± 0.07 mm), except the increase in abundance of buccal mucous cells.

5.2.2c Esophagus

At the mouth opening (8 DAH, TL = 12.85 ± 0.03 mm.), the buccopharynx communicated with the anterior intestine through a shorter and narrower esophagus, located directly posterior to the last gill-arch present in the pharynx. The anterior section of the incipient esophagus was filled with yolk and pigment granules until 10 DAH (Fig. 10A). With the gradual disappearance of yolk material between 10 and 12 DAH, the esophagus became completely pervious with the two different regions. The anterior esophageal region was lined by pseudostratified columnar epithelium cells, while the dilated posterior region was lined by a simple, cuboidal epithelium with abundant goblet mucosal cells and ciliated cells, from which the future stomach originates. On 10 DAH, several longitudinal folds were present at the junction between the posterior region of the esophagus and the stomach (Fig. 10B). These folds increased in complexity and length with larval age. Four tissue layers of the esophagus, namely mucosa (M), submucosa (S), muscularis (MU), and serosa (SE) also appeared on 12 DAH (Fig. 10B). From 12 DAH, the histological organization of the esophagus did not change except that the number and size of the goblet cells increased (Fig. 10C) in the posterior region of the esophagus and the longitudinal folds increased with a thicker muscular layer to accommodate the passage of large food items.

5.2.2d Stomach

At the beginning of hatching, the stomach was undifferentiated. At 10 DAH, the future stomach appeared as a little pocket with an esophagus-stomach sphincter.

The developing stomach was initially filled with yolk and pigment granules that transformed into the cylindrical epithelium, and the yolk material gradually disappeared. The first gastric glands became visible on 12 DAH, and their number increased through larval development. At 15 DAH (TL = 20.94 ± 0.06 mm.), the stomach exhibited a pouch shape and began to differentiate morphologically and histologically into three regions – the cardiac stomach, the stomach fundus, and the pyloric stomach. The wall of the stomach was very thin, and there were fewer mucosal folds in comparison with the esophagus and the intestine. In cross section, mucosae, submucosa, a muscular layer, and a serosa formed the stomach wall (Fig.11 A). Until 30 DAH, the stomach did not show any further noticeable modification except for the increase in the folds and goblet cells. The anterior non-glandular cardiac region, which continued at the end of the esophagus, was lined with a pseudostratified ciliated columnar epithelium with central nuclei, and it exhibited several mucosal folds and goblet cells. The mucosa of the stomach fundus was formed by a single layered columnar epithelium. The submucosa was formed by loose connective tissue containing blood vessels. The muscle consisted of smooth muscle fibers arranged in a circular inner layer and a thinner longitudinal outer layer. The epithelium of the glandular fundic region was similar to the cardiac region, but the lumen in this part appeared wider. The mucosa contained a large number of gastric glands on the dorsal and ventral walls, surrounded by the connective tissue of the lamina propria. The gastric glands were well established in the epithelium of this region. The eosinophilic particles of the gastric glands indicated pepsinogen secretion between 25 and 30 DAH (Fig. 11B). The wall of the pyloric portion comprised four layers, similarly to the stomach fundus, and it was marked by the sudden disappearance of gastric glands. With larval developmental advancement, the number and size of the mucosal folds increased.

5.2.2e Intestine

The intestine was the longest portion of the digestive tract and one of the first organs of the digestive system to differentiate in featherback larvae. At 4 DAH (TL = 10.65 ± 0.01 mm), the incipient intestine appeared as a straight translucent tubular segment laying dorsally to the yolk sac, and it was lined by a simple columnar epithelium. The intestinal segments showed a single epithelial layer composed of columnar cells with numerous microvilli at the luminal surface forming the brush border on 6 DAH (Fig. 12A). At this age, the intestinal cells, or enterocytes, were also arranged in a single layer, and contained median to basally located nuclei and evident apical vacuoles. With the coincidence of exogenous feeding commencement from 12 DAH, the lumen of the anterior intestine dilated and small, non-stained vacuoles could be seen in the cytoplasm of the intestinal enterocytes. As the larvae grew, the mucosal folds became deeper and more abundant, with an increase in number of intestinal mucus secreting goblet cells (Fig. 12 B and C). The increased thickness of the enterocyte brush border indicated an increase in digestion and absorption area. Eosinophilic supranuclear inclusion vesicles (SIV) in the posterior gut were first observed on 12 DAH (Fig. 12 D). SIV increased in frequency and size up to 15 DAH, but decreases in both the size and frequency of them occurred as development continued. The intestinal mucosa was surrounded by a thin musculature that comprised two muscular tissue layers: one circular internal and another longitudinal external separated by a very thin connective tissue layer. The intestinal mucosa showed a pronounced folding on 6 DAH, indicating the development of caecum, and the pyloric caeca joining the anterior intestine. The pyloric caeca became pronounced between 8-12 DAH and were different from the intestinal ones with their larger amounts of submucosa, and shorter folds. The lumen of the pyloric caeca was star like in cross section (Fig.12 E). The histological structure of the pyloric caeca was similar to that of the anterior intestine. In the distal portion, the folds containing large numbers of goblet cells were fewer and shorter, and they became progressively thicker. Intestinal mucosa developed longitudinal folds with the longest fold in the anterior section, which was shorter in the anterior- central part, and the shortest in the posterior intestine. The rectum was discernible as a short, flattened intestinal segment of the intestine devoid of mucosal folds (Fig. 12 F). Except for increases in the length of the intestine, as well as the size and number of mucosal folds, no relevant modifications were observed regarding the histological organization of the intestine until the end of the study.

5.2.2f Liver and Pancreas

Accessory glands (liver and pancreas) were not differentiated at hatching (Fig.13A). The primordial liver situated in the ventral region of the yolk sac was observed on 6 DAH (TL = 11.98 ± 0.05 mm), and it appeared histologically as a mass of polyhedral hepatocytes. These hepatic cells were filled with yolk, pigment granules, and large lipid vacuoles (Fig.13B) that were loosely organized around a central vein and hepatic sinusoids. On 8 DAH, liver size increased because of hepatocyte differentiation and the lipid vacuoles becoming more numerous between 6 to 8 DAH. On 10 DAH, a decrease in the lipid vacuoles was observed with a concurrent increase in eosinophilic granules (Fig. 13C). On the 8 DAH gallbladder and bile duct became visible but not fully pervious. They were lined with simple squamous epithelium, but later, on 10 DAH, three layers became notable in the gallbladder with a simple epithelium of columnar inner cell layer surrounded by a layer of connective tissue and smooth muscle fibers towards the outside part of the wall. At the same time (10 DAH), the hepatic yolk almost completely disappeared. Distinct histological changes took place after the onset of larval exogenous feeding. During this time, hepatocytes showed no more lipid vacuoles, and their shape became regular (Fig. 13D). Numerous blood vessels filled with blood cells and appeared among the hepatocytes. They were tightly packed between sinusoids, often around a central vein. Hepatocyte nuclei were located peripherally, and cytoplasm was almost completely filled with lipids. As the larvae grew, the livers continued to differentiate and liver size increased. Morphologically, it became a large organ that appeared ventral to the developing gut and formed a central anterior mass just behind the transverse septum, which separates the pericardial and the abdominal cavities. Between 6 and 8 DAH, the pancreas was first observed as small patches (diffused) of undifferentiated cells. With larval development, the pancreas started to move dorsally and compress around the stomach and the anterior intestine. On 8 DAH, exocrine polyhedral cells were visible with round basal nuclei arranged around a central lumen to form acini interspersed with blood vessels, containing dense eosinophilic zymogen granules at the apical portion of cytoplasm toward the lumen. The endocrine cells at 8 DAH could be distinguished as islets (islets of Langerhans) inside the exocrine pancreas (Fig. 13E).Pancreatic ducts were formed at this age, and the opening of the main pancreatic duct into the anterior intestine, just behind the pyloric sphincter, could be clearly seen at this age. The pancreatic duct was lined with squamous epithelium, which appeared flattened and thin. The histological organization of the pancreas did not show major changes from 12 DAH until the end of the study except in the increase in the size and number of pancreatic acini and the islets of Langerhans. With developmental advancement, mesenteric fat deposits observed surrounding the pancreatic tissue (Fig. 13F).

The major histological landmarks during the early development of *Chitala* is represented in Fig 14 and the significant histomorphological changes of the digestive tract are summarized in Table 3.

From the study it is evident that the featherback larvae depend exclusively on an endogenous source of nutrition during its first phase of life, i.e., from hatching until 8 DAH, when the yolk and oil globule reserves are metabolized. Between 8-12 DAH, the presence of yolk in the glandular stomach and anterior intestine indicated the mixed feeding period in *C. chitala*. From 12 DAH, with the complete exhaustion of the yolk, exogenous feeding commenced. The majority of histomorphological changes in the digestive system of *C. chitala* larvae occurred before the start of exogenous feeding. The development of the liver, pancreas, and ossified gill arches occurred before hatching. The mouth opens and absorptive and digestive capabilities are developed (i.e. enterocytes with microvilli, a pancreas with zymogen granules) develop before the start of exogenous feeding. These results concur with other altricial teleost larvae developing from demersal eggs which can be fed more easily on artificial diets very soon after hatching (Govoni et al., 1986). The lumen of the pharynx and esophagus of *C. chitala* was lined with numerous mucous cells secreting mucosubstances and ciliated cells which are adaptable for rapid and consistent lubrication during ingestion of small fishes, worms, crustaceans, and other small aquatic animals (Yang et al., 1997). The striated muscle fibers in the pharynx and esophagus allow for expansion during the ingestion of food (Albrecht et al., 2001). The development of teeth, taste buds, and goblet cells in the buccopharynx of *C. chitala* was similar to the pattern noted in other teleosts (Baglole et al.,1997; Papandroulakis et al., 2004), and all of these histological features play important roles in gestation and pregastric digestion (Sanchez-Amaya et al., 2007). The appearance of taste buds on the surface of the buccopharyngeal cavity at 10 DAH confirms the ability of *C. chitala* to forage for and recognize food (Sanchez-Amaya et al., 2007). It is remarkable that the pharynx of featherback larvae contains four gill arches with primordial gill filaments at 12 DAH, the development of gastric glands in fish larvae has been shown to indicate stomach differentiation (Verreth et al., 1992). In *C. chitala* larvae gastric gland development occurred on 12 DAH. It has been proposed that once the larvae complete metamorphosis, secretions of the gastric glands together with pepsin and HCl, facilitate the hydrolysis of proteins to peptides and amino acids (Govoni et al., 1986). Thus, the development of gastric glands helps larvae to digest a variety of prey items. This is particularly important in larval culture, as it indicates a potential dietary shift from live food organisms to artificial diets that can, consequently, reduce production costs. The location of gastric glands within the stomach is species-specific; an entirely glandular stomach, such as that found in the featherback larvae, suggests that this species could be capable of consuming large prey (Ortiz-Delgado et al., 2003). On approximately 6 DAH, pyloric caeca were seen first as slight evaginations of the anterior intestinal epithelium and the folding became distinct in the intestinal junction between 8 and 12 DAH, indicating improved digestion and absorption ability with the help of residing gut microflora and increases in the surface area (Baglole et al., 1997). The main function of the intestine, even in larvae, is the absorption of fat, protein, and carbohydrates. Possible functional capabilities can be inferred from the structural features of the differentiated cells composing the epithelium of the intestine (Ozaki, 1965). Numerous microvilli were observed at the intestinal luminal surface of *C. chitala* forming the brushborder. Welsch and Storch (1976) proposed that these structures are particularly very rich in enzymes to facilitate the digestive process. The presence of supranuclear inclusion vesicles (SIV) in the posterior gut of *C. chitala* are thought to be associated with pinocytotic absorption of protein macromolecules from the intestinal lumen into the enterocytes. It has been postulated that pinocytosis is an adaptation to the lack of a glandular stomach producing proteolytic enzymes during early development (Govoni et al., 1986; Hamlin et al., 2000). The SIV in the intestinal mucosa of *C. chitala* larvae were first observed at 12 DAH, i.e., at the onset of exogenous feeding, and then decreased in size and number around 15 DAH eventually disappearing with larval growth. The decrease in the size of SIV coincided with the proliferation of gastric glands at 15 DAH. Santamaria et al. (2004) correlated the disappearance of SIVs with increased enzymatic activity. However, SIVs were found to be absent in starved gilthead sea bream larvae (Yufera et al., 1996). The size and number of vacuoles decrease in consequence to the increased capacity of the enterocytes to synthesize lipoprotein (Deplano et al., 1991). The coexistence of pinocytosis and extracellular digestion in fish larvae has been described earlier and is considered to be an adaptive mechanism that compensates for the incomplete digestion of macromolecules until the proliferation of gastric glands (Govoni et al., 1986, Elbal et al., 2004). No relevant differences were observed in the histological organization of the anterior and posterior regions of the *C. chitala* larvae intestine, except in the number and size of mucosal folds and the degree of fat accumulation within the enterocytes. The intestinal goblet cells were the abundant cell type and were scattered along the intestinal epithelium of *C. chitala* larvae. These cells are well known for the production of mucins to establish a mucopolysaccharide barrier between the epithelium and the content of the lumen (Gisbert et al., 1998). Intestinal coiling and mucosal folding began at an early stage (10 DAH and 12 DAH, respectively) in *C. chitala* larvae and appeared very pronounced in the anterior intestine by 12 DAH, which indicated better gut functionality by increased intestinal length and absorption surface. The timing of liver and pancreas differentiation varies among species, and is mainly related to their general life history traits (Hoehne-Reitan and Kjørsvik, 2004). The early differentiation of the liver and pancreas in featherback larvae is considered as a glycogen storage, as eosinophilic granule in the liver, as a sign of the onset of hepatocyte functionality that is indicative of the synthesis and storage of macromolecules, which remain throughout the larval and juvenile stages. In the study, lipid accumulation appeared in the liver between 6 to 8 DAH. Thereafter, the level of fat deposits decreased and glycogen deposits were observed in hepatocyte cytoplasm which tend to store glycogen and lipid reserves in the liver as their digestive systems gradually develop and the pancreatic and intestinal functions are established (Zambonino-Infante et al., 2008). The histological development of the liver and bile transport system in *C. chitala* is thought to be concurrent with the gradual maturation of hepatocytes, as well as their functionality to synthesize, store, and mobilize carbohydrates and lipids (Hoehne-Reitan and Kjørsvik, 2004). The presence of the pancreas in *C. chitala* on 6 DAH prior to exogenous feeding confers the priority to accumulate and synthesize the pancreatic digestive enzymes for food digestion at the beginning of exogenous feeding (Sarasquete et al., 2001). The appearance of these pancreatic enzymatic precursors on 8 DAH prior to exogenous feeding supports these observations and suggests that the appearance of zymogen granules before feeding is an indication of a genetically programmed developmental process rather than a nutritionally induced event (Zambonino-Infante and Cahu, 2001; Micale et al., 2006).

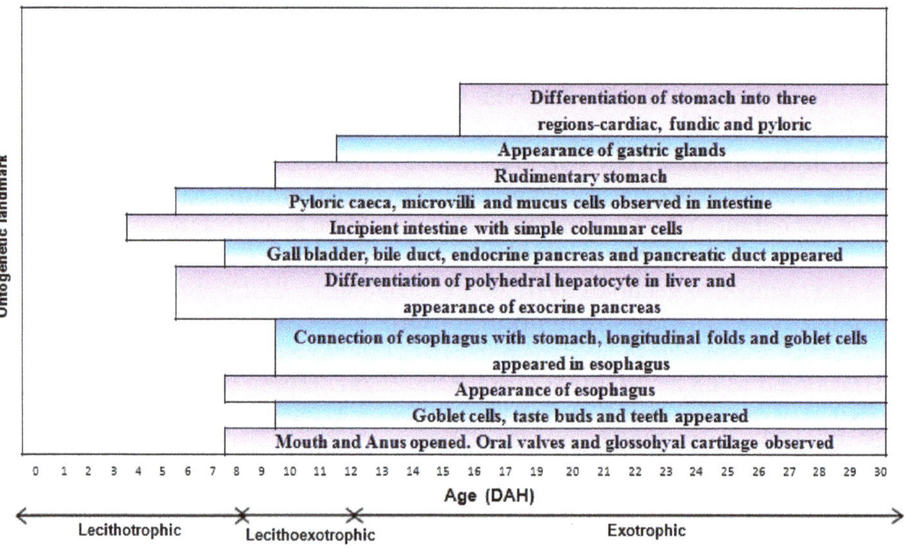

Fig 14: Diagram showing the main histological ontogenetic landmarks during the larval development of *C. chitala* (Mitra et al., 2015a)

Table 3: Summary of histological changes in the digestive system of *C. chitala* larvae from hatching to 30 days after hatching (DAH) (Mitra, 2015b)

DAH TL (mm)	Hatching 10.21 ± 0.03	4 10.65±0.01	6 11.98 ± 0.05	8 12.85 ± 0.03	10 16.9 ±0.05	12 16.9 ±0.05	15-30 20.94 ± 0.06-30.45 ± 0.07
Mouth, oral cavity, pharynx	Undifferentiated digestive tract lying dorsally to the yolk-sac without communication with the exterior			mouth opened, medial glossohyal cartilage, oral valves, primordial gill arches with pseudobranch	Canine like teeth, goblet cells	Taste bud goblet cells became numerous, gills with increased number of lamellae	increased in abundance of the buccal mucous cells
Esophagus				anterior section of incipient esophagus was filled with yolk and pigment granules	esophagus became completely pervious	Connection with stomach. Differentiation of secretory (with goblet cell) and food transport (longitudinal ciliated folds)\ regions. Four tissue layers of the oesophagus- namely mucosa, submucosa, muscularis and serosa	increased in the number and size of goblet cells in the posterior region of the esophagus and the longitudinal folds increased with a thicker muscular layer
Stomach					Rudimentary stomach appeared as a little pocket with an esophagus-stomach sphincter	first gastric glands became visible	Differentiated into three regions - the cardiac stomach, the stomach fundus and the pyloric stomach. In cross section, mucosae, submucosa, muscular layer and serosa formed the stomach wall. Increment of the folds and the goblet cells. eosinophilic particles of gastric glands indicated pepsinogen secretion

DAH TL (mm)	Hatching 10.21 ± 0.03	4 10.65±0.01	6 11.98 ± 0.05	8 12.85 ± 0.03	10 16.9 ±0.05	12 16.9 ±0.05	15-30 20.94 ± 0.06-30.45 ± 0.07
Intestine		primordial intestine. Simple columnar epithelium (enterocytes) Anus closed	numerous microvilli at the luminal surface forming the brush border pyloric caeca, rectum			Eosinophilic supranuclear inclusions vesicles (SIV) in the posterior gut were first observed	Intestine wall formed by mucosa, connective tissue, inner circular and outer longitudinal muscle layers increment in the length of the intestine, as well as the size and number of mucosal folds
Accessory glands			primordial liver with mass of polyhedral hepatocytes, small patches of undifferentiated pancreatic cells	gallbladder, bile duct and pancreatic duct became visible. exocrine polyhedral cells with round basal nuclei arranged to form acini interspersed with blood vessels, containing dense eosinophilic zymogen granules at the apical portion of cytoplasm towards lumen. Endocrine cells distinguished as islets of Langerhans inside exocrine pancreas	Hepatocytes (with large vacuoles) arranged along sinusoids	Increased in size of both organs	

5.3 Profile of Digestive Enzyme During Early Ontogeny

The success of effort to introduce and cultivate any new species depends on the adequate knowledge of nutritional requirement by understanding the functional development of their digestive system. The information on the appearance of key digestive enzymes during early ontogeny can be a good indicator of cultivable species in aquaculture and it is essential to develop age specific formulation of feeds that contributes to rapid and efficient growth rates. Mitra et al. (2015 b; 2016a) reported the development of key digestive enzymes in *Chitala* during early ontogeny. Following the similar protocol of fertilization and rearing of the larvae (Mitra et al., 2014b), the mouth opening and endoexotrophic feeding stage was observed at 8 DAH. But from 12 DAH the larvae exhibited complete exogenous feeding with complete absorption of yolk sac. Larvae were collected randomly from three tanks at a fixed time (9 a.m.) before morning feeding every other day in order to minimize the potential effects of exogenous enzymes from undigested live food in the gut. Larvae were euthanized with an overdose of tricaine methanesulfonate (MS-222, Sigma), rinsed in distilled water and after water removal with a filter paper. Then, samples were frozen in liquid nitrogen and stored at -20^0 C until further analysis. About 250 larvae per tank were collected and pooled on 1, 3, 5, 7 DAH; 100 larvae on 9, 12, 15,18 DAH and 50 larvae on 20, 25 and 30 DAH at each sampling time to determine the pattern of enzymatic activity. Following the collection, larvae were maintained in beakers filled with water for 1 h to allow any remaining food in the gut to be assimilated or excreted. Then, the larvae were frozen at -20^0 C until chemical analyses started. The frozen whole larvae were homogenized in four volumes (v/w) of ice-cold phosphate buffer at pH 7 using a hand held glass homogenizer followed by centrifugation at 10,000Xg for 20 min at 4^0 C. Supernatants or crude enzyme extract were stored at -20^0 C until analysis (Suzer et al., 2006). Soluble protein in the crude enzyme was measured by the Lowry method (Lowry et al., 1951) using bovine serum albumin as a standard. The a-amylase (EC.3.2.1.1) activity was determined in triplicate by using 1 % soluble starch, as substrate, with 3, 5-dinitrosalicyclic acid (DNS) at 550 nm (Bernfeld, 1951). The α-amylase activity was expressed as U = µg of maltose min^{-1} mg^{-1} of protein present in the enzyme extract tested. Pepsin activity was assayed using the method of Anson (1938) with bovine haemoglobin as the substrate. The pepsin (E.C.3.4.23.1) activity was quantified as U = µg of tyrosine min^{-1} mg^{-1} of protein present in the enzyme extract tested. Trypsin (E.C.3.4.21.4) activity was measured using the method of Hummel (1959) modified by Rick (1974a) with TAME (p toluenesulfonyl-L-arginine methyl ester) as the substrate. Trypsin activity was expressed as U = 0.001 ΔAbs247 min^{-1} mg^{-1} of protein present in the enzyme extract tested. Chymotrypsin (EC. 3.4.21.1) activity was determined using the method of Hummel (1959) described by Rick (1974b) with BTEE (N benzoyl- L-tyrosine ethyl ester) as the substrate. Specific chymotrypsin activity was measured as U = 0.001 ΔAbs256 min^{-1} mg^{-1} of protein present in the enzyme extract tested. Lipase (EC.3.1.1.3) activity was assayed based on measurement of fatty acids release due to enzymatic hydrolysis of triglycerides in stabilized emulsion of olive oil (Borlongan, 1990) and the activity of lipase is reported as U = µg of fatty acid min^{-1} mg^{-1} of protein present in the enzyme extract tested.

From the study it was observed that the α-amylase (i.e. from 1 DAH to 9 DAH) activity was initially high during endotrophic and endoexotrophic feeding stage. Since in the present experiment live food *(Artemia nauplii)* did not offer a suitable substrate for such an enzyme, substrate-mediated induction was not produced and the activity progressively decreased (Fig 15a and 15b). This suggested the synthesis of amylase in early ontogenesis even in absence of food, and the carbohydrates are actively catalyzed during the time of organogenesis. A constant decrease in activity after the 12 DAH i.e. on the onset of flexion and exotrophic stage of featherback fish may be possibly due to the developmental changes in the gut morphology (Chakrabarti et al., 2006a, b) and this pattern of increase in specific activity followed by a decrease to a constant level is characteristic of developing vertebrates and constituted a step in the metamorphosis (Cahu and Zambonino-Infante, 2001). According to Uscanga-Martínez et al. (2011), amylase has been found to be an integral component of the enzymatic equipment in developing larvae of carnivorous fish. The high α-amylase activity levels at late larval and juvenile stages in these carnivorous and omnivorous species would enable fish to more adequately take advantage of nutrients through the hydrolysis of glycogen from zooplanktonic and zoobenthic types of prey, when these species shift their feeding habits. It is commonly thought that lipase catalyzes the hydrolysis of emulsified esters of glycerol and long chain fatty acids. But beside this, lipase also has important role in carotenoid biosynthesis, transportation and pigmentation in fish (Gouveia and Rema, 2005). Lipase activity in endogenous feeding period or yolk sac phase of *C. chitala* might be related to the metabolism of lipid reserves in yolk accumulated before metamorphosis that were mobilized during the morphological modification or correspond to fatty acid requirements for larval development. But the increased activity during exotrophic phase (Fig 15c and 15d) can be correlated with the higher efficiency to digest and utilize the dietary lipid maximally. Changes in lipase activity in *C. chitala* might also be attributed to the production and secretion of pancreatic enzymes (Zambonino-Infante et al., 2009; Pradhan et al., 2012). During early ontogeny, before the onset of acidic digestion, protein digestion occurs mainly by the action of alkaline proteases such as trypsin and chymotrypsin in combination with intestinal cytosolic peptidases. During this period, these enzymes have limited capacity for digesting macromolecules that are absorbed by pinocytotic activity of the enterocytes in the posterior intestine for their intracellular digestion (Zambonino-Infante and Cahu, 2001). It is well noted that trypsin could be detected before mouth opening and exponentially increased by exogenous feeding, larval age and size coinciding with secretion of zymogen granules (Zambonino-Infante and Cahu, 2001; Zambonino-Infante et al., 2008). Chymotrypsin is an endopeptidase selectively hydrolyzes peptide bonds on the carboxyl side of the aromatic side chains of tyrosine, tryptophan and phenylalanine and also it is usually activated by trypsin (Zambonino-Infante and Cahu, 2001; Zambonino-Infante et al., 2008). It is presented similar profile during early developmental stage as well as the observed increase in specific activity and contributes to digestion of protein in larvae by partially compensating for the absence of acid proteases before the formation of a functional stomach (Applebaum et al., 2001; Zambonino- Infante and Cahu, 2001). It is well known that trypsin and chymotrypsin are the specific pancreatic protein-hydrolysing enzymes and belong to alkaline proteases with a considerable role in food digestion. The secretion of trypsin is known to occur in response to food ingestion and, in larval pancreatic tissue. In the study trypsin activity found at the hatching may come from the hatching gland in mother fish (trypsin-like hatching enzymes) (Cahu and Zambonino-Infante, 2001). The tryptic (Fig 15e and 15f) and chymotryptic (Fig 15g and 15h) activity was first detected in preflexion stage or endotrophic feeding stage from 1 DAH related to exogenous feeding in preflexion stage. The activity pattern of trypsin and chymotrypsin can be correlated by two explanations -first, the inherent enzymes available in the food may have caused an increase and second, enzymes production by the liver, pancreas and mucosal epithelium may have stimulated in response to initial food consumption and increased ration size with the development of *C. chitala*. The observed increase of chymotrypsin than the trypsin at an early developmental stage during the larval period of *C. chitala* suggests that chymotrypsin contributes more to protein digestion in larvae by partially compensating for the absence of acid proteases before the development of a true stomach. The increase in the trypsin and chymotrypsin activity immediately after the mouth opening (12–15 DAH) in the present study may be attributed to the transcriptional modulation for better protein digestion (Cahu and Zambonino-Infante, 2001). Pepsin is secreted by the gastric glands of the stomach, when present. Indeed, this organ is not differentiated at hatching and progressively develops during larval life. Hence, pepsin is expressed later than other digestive enzymes during the fish larval development. The absence of pepsin in the digestive tract of teleosts during early stages of life is thought to be compensated by micropinocytosis and intracellular digestion of proteins in the posterior intestine (Galaviz et al., 2012). The late detection of pepsin in the present study can be correlated with onset of the metamorphosis. The increase in specific pepsin activity (15 i and 15j) during exogenous feeding period or flexion stage onwards (15–30 DAH) had an effect on the reduction in serine protease (trypsin and chymotrypsin) activity on 18–30 and 20–30 DAH. The alkaline protease activity expressed as percentage of total proteolytic activity found in *C. chitala* larvae represented 100% of the total activity found in the guts on 8 DAH (Fig 16). This activity of alkaline protease decreased afterwards with the age. On 12 DAH the activity of acid proteases in relation to the total protease activity was only 39.68%. But it increased substantially by 18 DAH and represented 51% of the total protease activity measured. The relative activity of acidic protease was 59.57% at 30 DAH in comparison with the alkaline protease activity. The observed decrease in trypsin and chymotrypsin in the present experiment might be due to the increased acid digestion by pepsin activity and the maturity of digestive organ. Trypsin specific activity was relatively higher during early stages and then sharply and/or slightly decline was observed about after 20 DAH correlated with the development of gastric glands before acidic digestion in stomach (Moyano et al., 1996; Cara et al., 2003; Suzer et al., 2006, 2007; Gisbert et al., 2009). It is well evaluated that gastric secretion (pepsin and/or pepsin like activity) commonly accepted transition to acidic digestion from alkaline process, and also clear evident of formation of functional stomach. The decrease in the amylolytic, trypsin and chymotrypsin activities on 15 DAH onwards i.e. during flexion stage might be related to anatomical and physiological modifications before acquiring a constant activity towards adult mode of digestion and the polynomial trend of these enzyme activities may entail the adaptation of the fish towards the diet during ontogentic shift to post flexion juvenile stage (Chakrabarti et al., 2006a, b) with progressive appearance of the digestive organs and by the response to changes in the amount and composition of available food which is either genetically programmed or induced by substrate (Martinez et al., 1999). The changes in the C: N index,

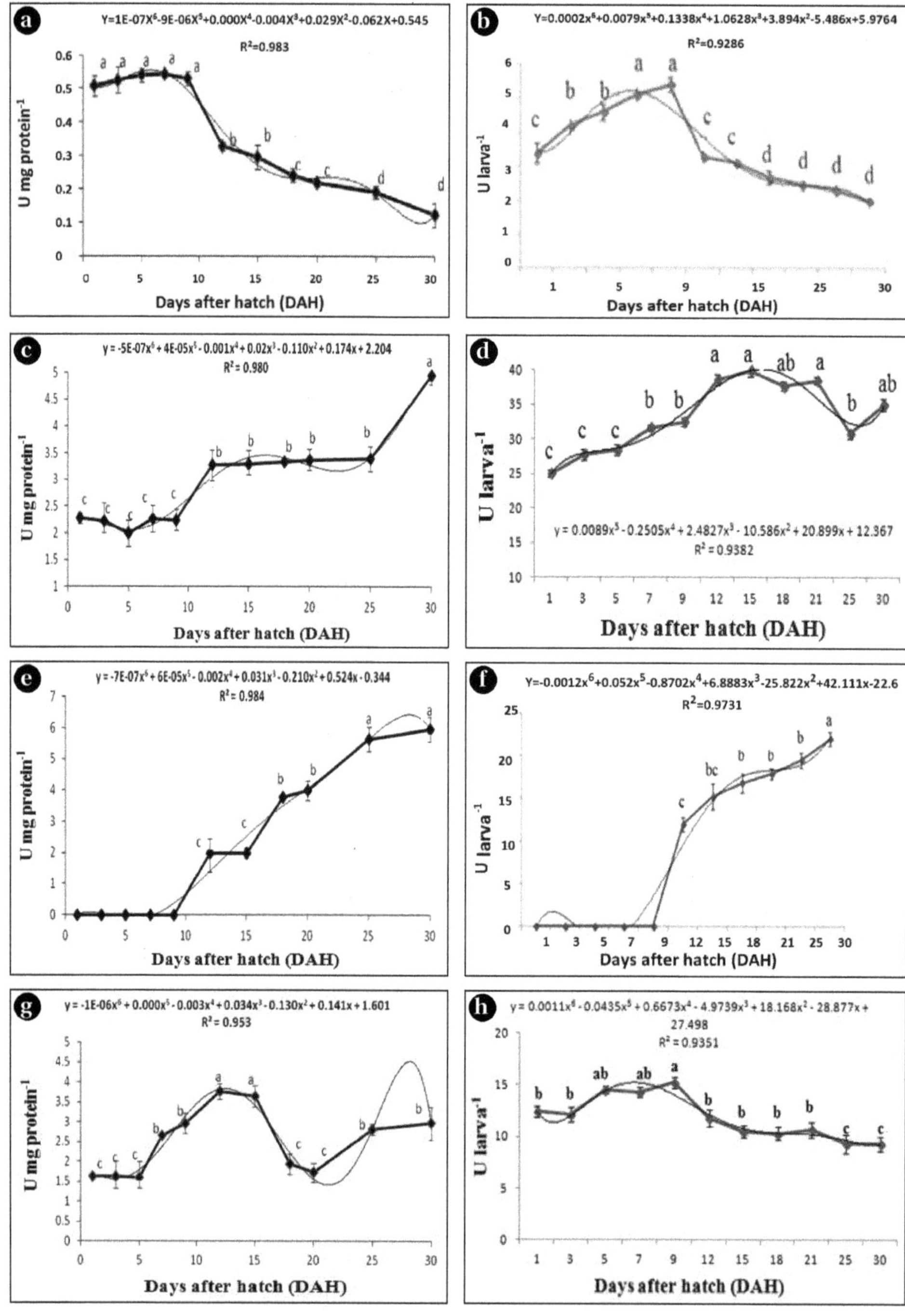

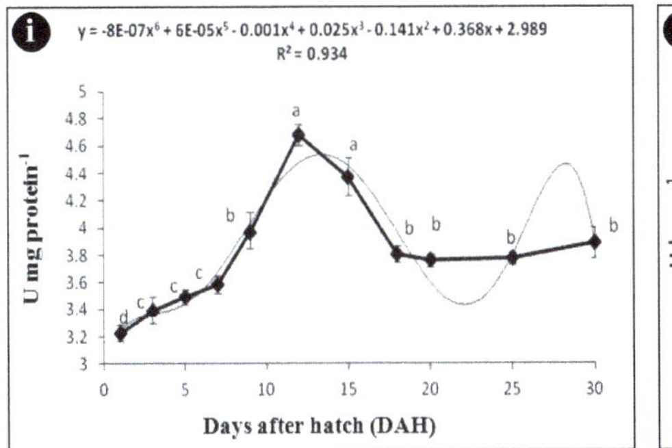

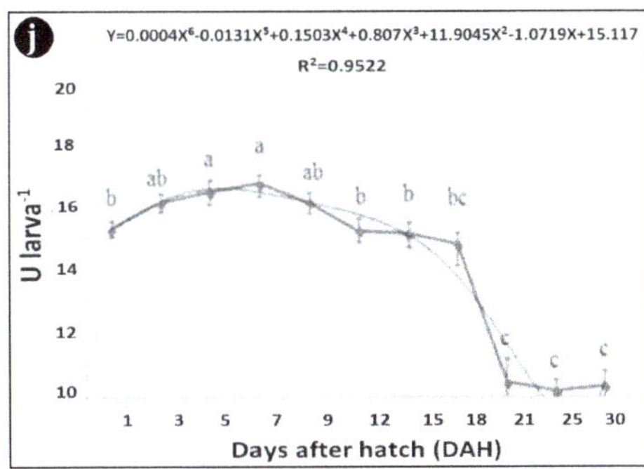

Fig 15: Specific (U mg^{-1} protein) and total (U larva^{-1}) enzymatic activities of α-amylase (a, b); lipase (c, d); trypsin (e, f); chymotrypsin (g, h); Pepsin (i, j) in *C. chitala* from hatching until 30 DAH (days after hatching). Results are expressed as mean ± SEM (n = 3). (Mitra et al., 2016a)

considered as an indicator of the ratio lipid: protein (Anger and Schultze, 1995) which reflects the existence of lipases able to hydrolyse lipids existing in the yolk sac. The ratio between lipase and protease activities in *C. chitala* during early development is shown in Fig.17. Fishes utilize lipids as the main source of energy at the first stages of development (Tocher and Sargent, 1984; Sargent et al., 1989), but progressively turn into a diet containing a greater proportion of protein. Following this hypothesis, relative proportions of lipase to total protease should decrease as the larva changes into a juvenile. As a general rule, the maximum activity of lipase was related to minimum protease activity (Cara et al., 2003). During the endogenous feeding period in *C. chitala* i.e. before 9 DAH the lipase: total protease ratio was >1.But at 9 DAH the ratio became 1:1 and thereafter the ratio of lipase: total protease found to be <1 until the experimental trial. The profile reflects the relative importance of lipid or protein metabolism of *C. chitala* larvae changed with time, and the lower relative activity of lipase may be correlated to depletion in reserves following yolk sac exhaustion (Mitra et al., 2015 b).

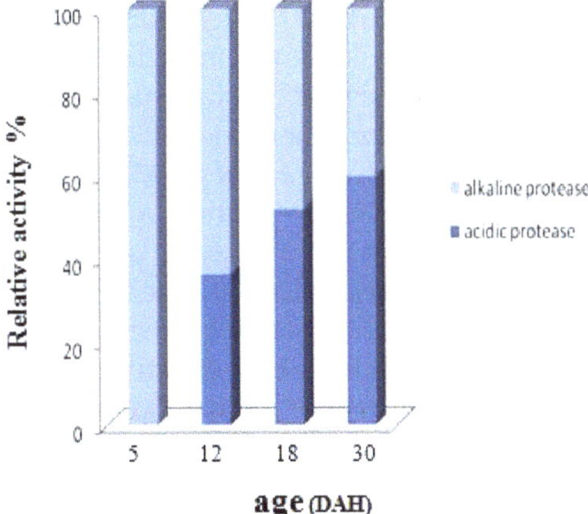

Fig 16: Relative activity % of acid and alkaline protease during different stages of *C. chitala* development (Mitra, 2015b)

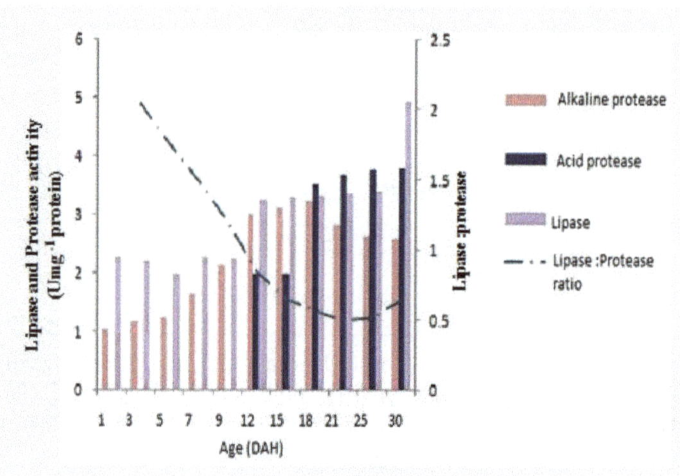

Fig 17 : Lipase, acid protease, alkaline protease and the ratio of lipase : protease activities during larval development of *C. chitala* larvae. (Mitra, 2015b)

5.4 Gut Microflora and Probiotic Approach for Nutritional Supplementation

5.4.1. Gut Microflora

The fishes have closer contact with the environmental microbiota due to their aqueous habitat and for this reason, transient microorganisms probably have more constant and important interaction with fish gastrointestinal ecosystems. The alimentary tract of fishes is a complex polymicrobial ecosystem and has a great influence on digestion and health. Fishes possess specific intestinal microbiota and the variation in the microbial flora in different fish species depends on the nutrition, intestinal microenvironment, age, geographical location, environmental factors, stress and etc. (Verschuere et al., 2000; Refstie et al., 2006; Skrodenyte-Arba iauskiene et al., 2008; Yang et al., 2007; Kesarcodi-Watson et al., 2008). The fish gut microflora have some specific roles which includes the metabolism of nutrients, contribution of the colonization resistance, antagonistic activity against pathogens and immunomodulation. Among the major group of microbes are represented in the fish gut, bacteria predominate (Spanggaard et al. 2000; Pond et al. 2006). Resident intestinal bacteria in fish are known to accelerate the digestion process by producing extracellular enzymes. The composition of enzyme producing bacterial flora in the fish digestive tracts can also be correlated to their feeding habits (Sugita et al., 1997). Thus understanding different aspects of microbial ecology of fish GI tract, presents a scope for fish nutritionists to formulate cost-effective fish probiotic supplements, especially in the growing stages (Bairagi et al., 2002). In this context Mitra et al. (2015b) identified the bacterial isolates from the intestine of the *C. chitala* fingerling and the types of isolates are represented in the Table: 4 with their morphological and biochemical features. The intestinal bacterial microflora isolated from *C. chitala* was diverse and consisted of *Aeromonas* spp., *Acinetobacter* spp., *Alcaligenes* spp.,

Bacillus spp., *Flavobacterium* spp., *Pseudomonas* spp., *Vibrio* spp. and *Enterobacter* spp. All of the isolates showed rod shaped structure. Among these, *Bacillus* spp. was only the gram positive motile bacteria. *Acinetobacter* spp. and *Flavobcterium* spp. was found to be the non motile gram negative isolates. *Vibrio* spp. was found to be the only isolate to produce H_2S. β-*hemolysis* was observed in case of *Aeromonas* spp., *Bacillus* spp., *Pseudomonas* spp. and *Vibrio* spp.

Characteristic Feature	Isolate I	Isolate II	Isolate III	Isolate IV	Isolate V	Isolate VI	Isolate VII	Isolate VIII
Gram Stain	-	-	-	+	-	-	-	-
Morphology	R	R	R	R	R	R	R	R
Motility	+	-	-	+	-	+	+	+
Indole	+	-	+	-	-	+	-	-
VP	+	-	-	-	-	+	-	+
Citrate	+	-	-	+	-	+	-	+
Methyl red test	+	-	+	+	-	-	-	-
H_2S Production	-	-	-	-	-	-	+	-
Oxidase	+	-	-	-	+	+	+	+
Urease	-	-	-	-	+	-	+	-
Catalase	+	+	-	+	+	+	+	-
Starch utilization	-	-	-	+	+	-	+	-
Acid from glucose	+	+	-	-	+	-	+	-
Gas from glucose	-	-	-	-	-	+	+	+
Hemolysis	β	-	-	β	-	β	β	-
Identified as	*Aeromonas*	*Acinetobacter*	*Alcaligenes*	*Bacillus*	*Flavobacterium*	*Pseudomonas*	*Vibrio*	*Enterobacter*

+,positive:-,negative:R,rod

Table 4: Physiological and Biochemical characteristics of Bacterial group isolates isolated from the intestine of juvenile featherback *C. chitala* (Mitra, 2015b)

The larger yellow colonies of *Bacillus* was selected and characterized (Mitra et al., 2014 a) on the basis of characters depicted in Table 5 and were identified using the guidelines described in the Manual of Determinative Bacteriology (Bergey, 1957). The spores were terminal and subterminal in position having ellipsoidal shape. All three organisms were rod shaped, short chained forming gram positive and capable of forming moderate colonies. Isolate I (*fb 9*) showed positive Methyl Red test (MR), positive starch hydrolysis, positive citrate utilization, positive casein hydrolysis negative H_2S production positive catalase, positive sugar fermentation with glucose, fructose, maltose. The isolate II (*fb 10*) was positive for MR test, citrate utilization and casein hydrolysis, glucose, fructose, xylose fermentation. Negative for nitrate reduction, starch hydrolysis and mannitol, sucrose, lactose and mannose fermentation. The isolate III (*fb 11*) showed positive to MR test, starch hydrolysis, citrate utilization, H_2S production, fructose, maltose sucrose, mannose, lactose fermentation.

The Bacillus isolates showed varied response under different physiological conditions. All the isolates showed optimum growth between 15-40^0 C. *Bacillus* isolate I (fb 9) showed optimum growth under pH 5-10. But in case of the isolates (fb 10 and *fb 11*) growth was observed under pH 2-9.However no growth was detected above pH 10. Isolate I (*fb 9*) and III (*fb 11*) was found to tolerate a wide range of NaCl concentration. These isolates were able to grow optimally under 2-9 % NaCl. The isolates (*fb 10*) and (*fb*) *11* had the optimal growth under 0.15-2% bile salt concentration (Table: 6).

Table 5: Morphological and Biochemical Tests for Bacillus Isolate I (*fb 9*), Isolate II (*fb 10*) and Isolate III (*fb 11*) isolated from juvenile *C. chitala* intestine (Mitra et al., 2014a)

Tests	Isolate I (*fb 9*)	Isolate II (fb 10)	Isolate III (*fb 11*)
Morphological tests			
Colony morphology:			
Configuration	Irregular, lobate	Round	Irregular
Elevation	Flat	Convex	Convex
Surface	Dull	Smooth	Rough
Pigments	ND	ND	ND
Gram's reaction	+	+	+
Shape	Rod shaped, short chained	Rod shaped, short chained	Rod shaped, short chained
Size	Moderate	Moderate	Moderate
Spore:			
Endospore	+	+	+
Shape, Position	Ellipsoidal, subterminal	Ellipsoidal, terminal	Ellipsoidal, subterminal
Biochemical tests			
Indole production	ND	ND	ND
Methyl Red Test	+	+	+
Citrate Utilization	+	+	+
Casein Hydrolysis	+	+	+
Starch Hydrolysis	+	ND	+
Urea Hydrolysis	ND	ND	ND
Nitrate Reduction	+	ND	ND

+, positive; ND, not detected

Table 6: Physiological Tests for *Bacillus* Isolate I (*fb 9*), Isolate II (*fb 10*) and Isolate III (*fb 11*) isolated from juvenile *C. chitala* intestine to understand the probiotic potential (Mitra, 2015b)

Tests	Isolate I (*fb 9*)	Isolate II (*fb 10*)	Isolate III (*fb 11*)
Growth temperature			
4°C	ND	ND	ND
15-40°C	+	+	+
55°C	ND	ND	ND

Tests	Isolate I (*fb 9*)	Isolate II (*fb 10*)	Isolate III (*fb 11*)
Growth at pH			
2-5	ND	+	+
5-9	+	+	+
10	+	ND	ND
NaCl (%) tolerance			
2-9	+	+	+
10	+	ND	+
Bile salt tolerance(%)			
0.15-0.3	+	+	+
2	ND	+	+

+, positive; ND, not detected

Extracellular enzyme production by bacterial strains isolated from the digestive tract of the investigated fish was assayed qualitatively (Table:7).The isolates were found to produce varied range of extracellular enzymes such as amylase, protease and lipase. The α-amylase activity was found to be the high (10-12 mm halo diameter) in fb 9 and the amylolytic activity was absent in *fb 10*. The fb 11 isolate showed very high protease producing ability followed by *fb 10* and *fb 9* isolate. But the isolates *fb 10* and *fb 9* showed moderate lipolytic activity (7-9 mm halo diameter) among the three isolates with the very high activity in case of isolate *fb 11*. From the invitro studies of enzyme producing capacity it was evident that the bacterial flora present in the gastrointestinal tract of *C. chitala* are good producer of proteolytic and lipolytic enzymes with a moderate producer of amylolytic enzyme and this can be correlated with the feeding habit of *C. chitala*, which is commonly carnivore and insectivore in nature (Alikunhi, 1957).

Table 7: Qualitative extracellular enzyme production by the Bacillus Isolate I (*fb 9*), Isolate II (*fb 10*) and Isolate III (*fb 11*) isolated from *C. chitala* gut (Mitra, 2015b)

Enzymatic activity	Isolate I (*fb 9*)	Isolate II (*fb 10*)	Isolate III (*fb 11*)
Amylase	+++	absent	+
Protease	+	+++	++++
Lipase	++	++	++++

++++, very high (13-16 mm halo diameter); +++, high (10-12 mm halo diameter); ++, moderate (7-9 mm halo diameter); +, low (4-6 mm halo diameter)

5.4.2 Probiotic Supplementation

The use of probiotic as feed additives has been proven to be used to improve feed utilization efficiency and health conditions during the early life stages of fish rearing. The Gram-positive, spore forming bacteria, *Bacillus* spp. are most widely used as commercial probiotics and positively enhance the survival and growth of the cultured organisms (Gomez-Gil et al., 2000). *Bacillus* enzymes are very efficient in breaking down a large variety of carbohydrates, lipids and proteins into smaller units and they can be grown efficiently with very low-cost carbon and nitrogen sources and it also has long lasting shelf life (Sonnenschein et al., 1993). They are able to exhibit competitive exclusion of pathogenic bacteria via producing inhibitory compounds (Verschuere et al., 2000). Taking into account these advantageous characteristics of *Bacillus,* Mitra (2015b) formulated five isocaloric (430 Kcal100g^{-1}), isonitrogenous (40% CP) experimental diets with different concentrations of B. licheniformis fb 11 probionts (isolated from the gut of *Chitala chitala*) viz. Control (without probionts), 5x10^{4} CFU g^{-1} (D_1), 5X10^{5} CFU g^{-1} (D_2), 5 x10^{6} CFUg^{-1} (D_3) 5 x 10^{7} CFUg^{-1} (D_4), 5 x 108 CFU g^{-1} (D_5) to evaluate its efficiency on growth performance, digestive enzyme activity, effect on gut microbiota and water quality in *C. chitala* juvenile.

To perform the experiment, a batch of healthy fingerlings of *C. chitala* (avg wt.3.5±0.75g) of a 2 years old brood stock (weight=2.31±0.6 kg, total length=86±5.5cm) collected from a local fish farm in West Bengal, India. The fishes were introduced to a growth tank equipped to hold them during an acclimation period of 15 days under the growth conditions of 28^{0} C, pH 7.5 mg L^{-1} dissolved oxygen. During this period the fishes were fed with the formulated basal diet other than the probiotic experimental diets. The uniform sized healthy juveniles were equally distributed in six dietary treatment groups with three replicates each (stocking density of fifteen fishes) per aquarium (20"x10"x12") .The feeding trial continued for a period of 90 days under laboratory conditions with continuous aeration by an air compressor. The photoperiod was maintained as 12 D:12 L with a constant water flow rate of 0.6 L min^{-1}. All the fishes were fed thrice a day at 9.00 AM, 1.00 P.M. and at 5.00 PM the feeding rate being 3% of the total body weight per day. The daily ration was adjusted every ten days on the basis of the weight increment. Uneaten feed was removed and stored separately to calculate the feed conversion ratio. Fecal samples were collected by pipetting (Spyridakis et al., 1989). The fecal samples were oven dried (60^{0} C) and analyzed for digestibility determinations. The proximate composition for experimental diets, faecal matter and whole body fish were measured according to AOAC (2000). The proximate analyses of the carcasses were done before initiation and after termination of the experiment following the procedures used for the diets. The apparent digestibility coefficient of dry matter (ADC_{DM}) was calculated as: 100-(% marker in feed/% marker in faeces x 100) and the apparent digestibility (AD) of nutrients (protein and lipid) were calculated as: 100 – 100 x (% marker in diet/% marker in faeces) x (% nutrient in faeces/% nutrient in diet).

Weights of all the fishes from each aquarium were determined at the beginning and end of the 90-days experiment, considered as initial weight and final weight, respectively. The growth performance was calculated using mathematical growth models as followed:

$\%\text{weight gain}=[(W_f-W_i)/W_i]\times100$

$SGR\ (\%/day) = 100\times[(L_nW_f - L_nW_i)/T]$

$FCR = TFI/(FB - IB)$

FCE (%) = [(FB – IB)/TFI]×100

CF = 100×(W/TL^3)

TGC (%) = [FB 0.333–IB 0.333/Σ°C (day-degrees)]×100

ADG = 100×[(W_f – W_i)/(W_i)×T]

RFI = 100×[(TFI)/0.5(W_f – W_i)×T]×100

PER = (FB – IB)/PI

LER= (FB – IB)/LI

where: SGR = specific growth rate; FCR = feed conversion ratio; FCE = food conversion efficiency; CF = condition factor; TGC = thermal growth coefficient; ADG = average daily growth; RFI = relative food intake; PER = protein efficiency ratio; LER = lipid efficiency ratio; W_f and W_i = final and initial body weight (g); T = time of rearing (days); TL = total length (cm); FB and IB = final and initial fish biomass (g); TFI = total feed intake (g), Σ°C (day-degrees) = daily mean temperature; PI = protein intake (g); LI = lipid intake (g).

The result of the study indicated that a better growth performances and feed utilisation (Table 8 and 9) was observed in case of the featherback juvenile fed with the probiotic supplemented diets at the concentration of 5×10^6 CFU 100 g^{-1}. The ability of *Bacillus* to improve the intestinal microbial balance by synthesizing the essential nutrients (vitamins and short chain fatty acids) and enzymes (amylase, protease and lipase), resulted into the better growth performance of the host (Gatesoupe, 1999). The highest crude protein was recorded for the fish fed with diet supplemented at the concentration of 5×10^6 CFUg^{-1} while the lowest crude protein observed from the fish fed with control diet. The results clearly indicate that the bacterial enzyme (protease, in particular) helped for better protein utilization, resulting in more protein accretion in the muscle. The protein and lipid digestibility was higher with the probiotic up to the supplementation 5×10^6 CFUg^{-1} and was positively correlated to the growth performance of the fish. But with the higher inclusion rate of bacteria, the digestibility gradually decreased. The less body fat content of fish carcass indicated the required energy for metabolism in *C. chitala* juvenile might be primarily supplied by fat and hence induced a reduction of lipid deposition with an increase in body protein. This can be considered as an important achievement in the production of fish with more protein and less fat.

For enzymatic analysis, fishes starved for 24 h were collected on days 0, 15, 30, 45, 60, 75 and 90 days of the trials and anesthetised in 0.02% MS-222. By using an ice plate, the intestine was dissected, rinsed with cold deionised water and immediately stored at -20^0 C before use. Digestive enzymes were extracted by homogenizing the intestine in ice-cold 0.1 M phosphate buffer at pH 7.0 using a hand held glass homogenizer followed by centrifugation at 10,000 g for 20 min at 4^{0}C. The supernatant was used for enzyme assay. Protein content of the extract was assayed according to Lowry et al. (1951). Specific activity of α- amylase was determined using 1% starch as the substrate (Bernfeld, 1951) and the specific -amylase activity is reported as U = μg of maltose min^{-1} mg^{-1} of protein present in the enzyme extract tested. Total protease activity was evaluated using the method of Garcia-Carreno (1992) using casein (0.5%) as the substrate. Specific protease activity is reported as U = µg of tyrosine min^{-1} mg^{-1} of protein present in the enzyme extract tested and the lipase activity was determined by using olive oil as the substrate (Borlongan, 1990). The specific activity of lipase is reported as U= µg of fatty acid $min^{-1}mg^{-1}$of protein present in the enzyme extract tested. A significant difference in digestive enzymatic activities (Table 10) of *C. chitala* juvenile was detected between basal diet fed and *Bacillus licheniformis fb11* live supplemented group. This can be correlated with the improved digestibility and feed utilisation with the help of endogenous enzymes secreted from the superiorly matured intestinal cells stimulated by the wide range of exozymes from *Bacillus* spp. (Munilla-Moran et al., 1990; Ziaei-Nejad et al., 2006).

The gastrointestinal micro flora analysis was carried out on 0th, 15th, 30th, 45th, 60th, 75th and 90th day of trial. The gut homogenates were serially diluted by NSS (Qualigens) and spread plated in triplicate onto TSA(Hi-Media), Pseudomonas Isolation Agar (Hi-Media) and Starch Ampicillin Agar (Palumbo, 1985) plates and incubated at 35^0 C for 48 h for the enumeration of total heterotrophic count (THC) and Bacillus spp., presumptive pseudomonads count (PPC) and motile aeromonads count (MAC) respectively. The total coliforms count (TCC) in gut samples were determined presumptively in lactose broth (Hi-Media) by Most Probable Number (MPN) – 5 tube technique followed by confirmatory testing in Brilliant green lactose bile (BGLB) broth (Hi-Media) with incubation at 35^0 C for 48 h (APHA, 1992). A significant increase in the total heterotrophic bacterial count (THC) and Bacillus spp. with concurrent decrease in the count of gut Pseudomonas, Aeromonas and Coliform was observed in *C. chitala* juvenile fed with the probiotic supplemented diet (Table 11). The reason may be attributed to the competitive exclusion by the *B. licheniformis fb 11* producing the antibiotics (polymyxin, bacitracin and gramicidin) and colonizing with the help of specific mucus adherence and cell surface hydrophobicity (Balcázar et al., 2006). It was also observed during the study that the probiotic *Bacilli* count in the feed reduced to roughly about 1 log unit of storage at 4^0 C after 90 days of storage (Table 12). The initial high viability of *Bacilli* in the feed was due to its endospores forming ability but the reduced (Gatesoupe, 1999) activity after 90 days of storage was might be due to different factors such as storage temperature, moisture content, chemical composition of the feed (Peighambardousta et al., 2011). Thus from the management planning aspect of the probiotic-supplemented feed application in fish culture, it is advisable to use freshly prepared probiotic feeds as the viability count of probiotic bacteria is reduced in the diets due to storage. As we all know that water quality plays important role in growth and survival of aquatic organisms. So from the culture perspectives it is very important to analyse the water parameters after probiotic supplementation in the diet of the fishes. Hence the physicochemical characteristic of water, viz., temperature, pH, dissolved oxygen, free carbon di oxide, total alkalinity, total ammonia were analyzed weekly till the end of the experiment following the methods outlined by APHA (2005). But during the entire experimental period and at the end of the experiment all the water quality parameters were found to be in the optimum range (Kumar et al., 2010; Mohapatra et al., 2012) of fish rearing and had no adverse effect on water quality (Table 13).

Table 8 : Growth performance, apparent digestibility of *Chitala chitala* juvenile fed with diets of different levels of probiotic supplements (Mitra, 2015b)

Growth Parameter	Treatments					
	D_0	D_1	D_2	D_3	D_4	D_5
%weight gain	246.21±12.11^{c}	261±11.6^{b}	296.69±12.28^{a}	305.43±19.75^{a}	286.12±14.22^{ab}	266.11±11.96^{b}
SGR	0.6±0.03^{b}	0.63±0.05^{b}	0.71±0.02^{a}	0.82±0.05^{a}	0.68±0.03^{ab}	0.62±0.01^{b}
FCR	1.51±0.02^{c}	1.65±0.08^{b}	1.72±0.05^{ab}	1.86±0.07^{a}	1.74±0.09^{ab}	1.61±0.03^{b}
FCE	57±1.5^{c}	63.1±2.03^{b}	68.21±1.56^{ab}	70.14±1.81^{a}	64.32±1.09^{b}	60.29±1.67^{c}
CF	1.18±0.08	1.18±0.03	1.21±0.025	1.19±0.05	1.20±0.02	1.19±0.015
TGC	0.67±0.05	0.68±0.03	0.67±0.02	0.70±0.05	0.68±0.01	0.68±0.03
ADG	2.85±0.06^{c}	2.07±0.09^{b}	3.15±0.03^{a}	3.31±0.06^{a}	3.06±0.08^{ab}	2.65±0.05^{b}
RFI	15.17±0.4^{a}	15.16±0.29^{a}	15.08±0.78^{a}	15.15±0.64^{a}	13.82±0.25^{b}	13.64±0.2^{b}
PER	2.20±0.09^{c}	2.28±0.06^{b}	2.34±0.08^{ab}	2.61±0.09^{a}	2.37±0.25^{ab}	2.12±0.08^{c}
LER	2.12±0.05^{d}	2.26±0.07^{c}	2.24±0.05^{c}	2.45±0.01^{a}	2.31±0.09^{b}	2.19±0.02^{d}
ADC_{DM}	49.1±0.74^{b}	51.21±0.25^{ab}	52.16±0.82^{ab}	53.8±0.68^{a}	53.28±0.9^{a}	49.69±0.8^{b}
$ADC_{Protein}$	75.7±0.52^{d}	81.26±0.87^{b}	84.12±0.52^{ab}	86.7±0.78^{a}	82.15±0.23^{b}	77.8±0.85^{c}
ADC_{Lipid}	73.34±0.26^{d}	77.52±0.43^{c}	78.28±0.38^{c}	83.4±0.5^{a}	81.27±0.75^{b}	79.5±0.78^{c}
Survival %	90.3±0.5^{b}	97.3±0.2^{a}	98±0.25^{a}	98.5±0.5^{a}	97.6±0.25^{a}	97±0.7^{a}

Values (Mean ±SD) with the different superscript differ significantly ($P<0.05$)

Table 9: Whole body chemical composition (g $100g^{-1}$ on dry matter basis) of fish in different probiotic supplemented groups (Mitra, 2015b)

Parameters	Treatments					
	D_0	D_1	D_2	D_3	D_4	D_5
Moisture	78.8±1.6	78.3±1.5	77.2±1.4	78.6±0.1	78.4±1.5	77.6±1.2
Total ash	18.21±0.5^{a}	18.20±0.29^{a}	16.10±0.35^{b}	15.66±0.52^{c}	18.17±0.21^{a}	16.56±0.42^{b}
Crude protein	56.32±0.45^{c}	58.14±0.7^{b}	61.83±0.61^{a}	63.78±0.45^{a}	59.16±0.34^{ab}	58.87±0.27^{b}
Ether extract	12.09±0.5^{a}	11.7±0.75^{b}	11.49±0.27^{b}	11. 41±0.63^{c}	12.39±0.58^{a}	12.12±0.4^{a}
Total carbohydrate	13.38±0.2^{a}	11.96±0.7^{bc}	10.58±0.26^{c}	9.15±0.31^{d}	10.28±0.26^{c}	12.45±0.28^{b}
Gross Energy (MJ $100g^{-1}$)	18.06±0.38^{b}	18.19±0.29^{b}	17.98±0.28^{c}	18.75±0.21^{a}	18.49±0.33^{a}	17.96±0.15^{c}

Values (Mean ±SD) with the different superscript differ significantly ($P<0.05$).

Table 10: Digestive enzyme activities in different doses of dietary probiotic supplemented diet on *C. chitala* at various sampling days (Mitra, 2015b)

Treatments		Days of Treatments						
		0	15	30	45	60	75	90
α-Amylase activity	**D_0**	0.23±0.008^{Aa}	0.24±0.004^{Ab}	0.28±0.005^{Ab}	0.26±0.006^{Ab}	0.26±0.004^{Ac}	0.23±0.005^{Ac}	0.26±0.005^{Ac}
	D_1	0.2±0.004^{Ca}	0.29±0.005^{Bab}	0.38±0.002^{AB}	0.31±0.004^{Aab}	0.35±0.005^{A}	0.3±0.004^{ABbc}	0.23±0.007^{Cc}
	D_2	0.21±0.005^{Ca}	0.31±0.0075^{Ba}	0.41±0.007^{Aa}	0.42±0.004^{Aa}	0.39±0.001^{ABab}	0.32±0.006^{Bb}	0.28±0.006^{BCb}
	D_3	0.2±0.003^{Ca}	0.35±0.004^{Ba}	0.43±0.006^{ABa}	0.47±0.003^{ABa}	0.5±0.004^{Aa}	0.48±0.006^{Aa}	0.31±0.001^{Ba}
	D_4	0.22±0.008^{Ca}	0.3±0.007^{ABab}	0.38±0.005^{Aab}	0.36±0.005^{Aab}	0.3±0.005^{ABb}	0.26±0.007^{Bbc}	0.24±0.006^{Cc}
	D_5	0.2±0.005^{Ca}	0.32±0.006^{Aa}	0.38±0.006^{Aab}	0.36±0.004^{Aab}	0.38±0.007^{Aab}	0.23±0.006^{Cc}	0.28±0.0055^{Bb}
Total Protease	**D_0**	2.5±0.21^{Aa}	2.9±0.3^{Ab}	3.1±0.2^{Ac}	3.1±0.25^{Ab}	3.21±0.22^{Ac}	3.2±0.27^{Ac}	2.8±0.26^{Ab}
	D_1	2.8±0.23^{Ca}	^{3}.1±0.25^{Bab}	3.9±0.23^{Aab}	4.1±0.28^{Aa}	3.7±0.21^{ABb}	3.5±0.24^{ABbc}	3±0.25^{Bb}
	D_2	2.8±0.18^{Ca}	3.2±0.2^{Bab}	4±0.34^{Aab}	3.8±0.2^{ABab}	4.2±0.3^{Ab}	3.9±0.22^{Ab}	3.6±0.2^{ABa}
	D_3	2.6±0.2^{Ca}	3.9±0.22^{Ba}	4.6±0.3^{ABa}	4.8±0.29^{Aa}	5.2±0.33^{Aa}	4.9±0.3^{Aa}	3.2±0.23^{BCab}
	D_4	2.65±0.23^{Ca}	3±0.2^{Bb}	3.2±0.28^{ABc}	3.8±0.25^{Aab}	3.42±0.26^{ABbc}	3.6±0.2^{Abc}	3.2±0.25^{Bab}
	D_5	2.72±0.2^{Ca}	3.6±0.31^{Aa}	3.7±0.2^{Ab}	3.57±.0.2^{Ab}	3.48±0.2^{ABbc}	3±0.27^{ABc}	2.9±0.2^{Bb}
Lipase activity	D_0	1.1±0.06^{Aa}	1.15±0.04^{Ac}	1.18±0.07^{Ac}	1.12±0.025^{Ac}	1.29±0.05^{Ac}	1.26±0.044^{Ac}	1.25±0.05^{Ac}
	D_1	1.25±0.62^{Ca}	1.8±0.06^{Bb}	2.04±0.02^{ABbc}	2.12±0.05^{Aab}	2.16±0.045^{Aab}	2.09±0.08^{Aab}	1.9±0.015^{Bab}
	D_2	1.3±0.05^{Ca}	2.09±0.08^{ABb}	2.21±0.034^{Aab}	2.16±0.046^{Bab}	1.89±0.067^{Bb}	1.95±0.05^{Bab}	1.62±0.01^{Bb}
	D_3	1.27±0.058^{Ca}	2.16±0.05^{Ba}	2.34±0.025^{Ba}	2.51±0.075^{Aa}	2.56±0.07^{Aa}	2.58±0.064^{Aa}	2.09±0.03^{Ba}
	D_4	1.3±0.06^{Ca}	2.07±0.01^{Ab}	2.16±0.06^{Ab}	1.58±0.02^{ABbc}	1.51±0.03^{Bbc}	1.43±0.023^{AB}	1.4±0.058^{Cbc}
	D_5	1.31±0.05^{Ca}	2.11±0.075^{Aab}	2.11±0.04^{Ab}	1.79±0.025^{ABb}	1.44±0.08^{Bbc}	1.39±0.04^{Bb}	1.36±0.075^{Cbc}

Values (Mean ± SD) with different upper case superscripts denote significant difference significant difference during experimental period ($P<0.05$), and values (Mean ± SD) with different lower case superscripts denote among the treatments ($P<0.05$)

Table 11: Effect of dietary supplementation of *B. licheniformis* fE11 on the gut microbiota of *C. chitala* at various sampling days in different dietary probiotic supplemented diet (Mitra, 2015b)

Treatments		Mean number of bacteria in log_{10} cfu g^{-1} gut tissue						
		0th day	15th day	30th day	45th day	60th day	75th day	90th day
THB (Total heterotrophic bacteria)	D_0	6.69 ± 0.15^{Aa}	6.73 ± 0.13^{Aa}	7.09 ± 0.16^{Aab}	7.14 ± 0.11^{Ab}	7.41 ± 0.14^{Ab}	7.64 ± 0.09^{Abc}	7.71 ± 0.11^{Ac}
	D_1	6.66 ± 0.16^{Aa}	7.19 ± 0.15^{Bb}	7.33 ± 0.11^{Abc}	7.37 ± 0.19^{ABbc}	8.27 ± 0.12^{Bc}	8.32 ± 0.17^{Bc}	8.45 ± 0.16^{Bd}
	D_2	6.72 ± 0.16^{Aa}	7.28 ± 0.12^{Bb}	7.42 ± 0.14^{Abc}	7.54 ± 0.11^{ABbc}	8.43 ± 0.17^{BCc}	8.45 ± 0.2^{Bc}	8.49 ± 0.21^{Bc}
	D_3	6.58 ± 0.11^{Aa}	7.39 ± 0.22^{BCb}	7.45 ± 0.12^{Ab}	7.72 ± 0.16^{Bbc}	8.52 ± 0.12^{BCc}	8.69 ± 0.22^{BCd}	8.75 ± 0.19^{Cd}
	D_4	6.67 ± 0.17^{Aa}	7.65 ± 0.11^{Cb}	7.77 ± 0.1^{Ab}	7.81 ± 0.11^{Bb}	8.57 ± 0.11^{Cbc}	8.62 ± 0.14^{BCbc}	8.7 ± 0.17^{BCc}
	D_5	6.61 ± 0.15^{Aa}	7.72 ± 0.19^{Cb}	7.81 ± 0.11^{Ab}	7.86 ± 0.12^{Bb}	8.61 ± 0.1^{Cc}	8.64 ± 0.1^{Cc}	8.67 ± 0.11^{BCc}
***Bacillus* spp.**	D_0	1.11 ± 0.07^{Aa}	1.19 ± 0.09^{Aa}	1.25 ± 0.02^{Aa}	1.32 ± 0.07^{Aab}	1.36 ± 0.02^{Ab}	1.42 ± 0.04^{Ab}	1.49 ± 0.08^{Ab}
	D_1	1.08 ± 0.11^{Aa}	2.48 ± 0.05^{Bb}	2.56 ± 0.08^{Bb}	2.62 ± 0.02^{Bb}	4.17 ± 0.01^{Bc}	4.28 ± 0.05^{Bc}	4.32 ± 0.03^{Bc}
	D_2	1.15 ± 0.09^{Aa}	3.21 ± 0.08^{BCb}	3.23 ± 0.11^{BCb}	3.26 ± 0.05^{BCb}	4.37 ± 0.08^{BCc}	4.57 ± 0.01^{BCd}	4.61 ± 0.02^{BCd}
	D_3	1.15 ± 0.01^{Aa}	3.54 ± 0.11^{Cb}	3.57 ± 0.05^{Cb}	3.6 ± 0.04^{Cb}	4.44 ± 0.11^{BCc}	4.68 ± 0.06^{Ccd}	4.79 ± 0.01^{Cd}
	D_4	1.11 ± 0.05^{Aa}	3.62 ± 0.08^{Cb}	3.69 ± 0.02^{Cb}	3.73 ± 0.02^{Db}	4.56 ± 0.04^{Cc}	4.69 ± 0.08^{Cd}	4.7 ± 0.09^{Bd}
	D_5	1.09 ± 0.08^{Aa}	3.68 ± 0.03^{Cb}	3.69 ± 0.11^{Cb}	3.78 ± 0.01^{Db}	4.61 ± 0.08^{Cbc}	4.74 ± 0.04^{Dc}	4.76 ± 0.06^{Cc}
MAC (Motile Aeromonad count)	D_0	3.67 ± 0.14^{Aa}	3.71 ± 0.11^{Aa}	3.72 ± 0.08^{Aa}	3.74 ± 0.02^{Ca}	3.79 ± 0.01^{Cab}	3.82 ± 0.01^{Dab}	3.86 ± 0.09^{Cb}
	D_1	3.68 ± 0.07^{Ab}	3.63 ± 0.04^{Ab}	3.6 ± 0.05^{Aab}	2.78 ± 0.05^{Ba}	2.73 ± 0.01^{Ba}	2.68 ± 0.06^{Ba}	2.64 ± 0.01^{Ba}
	D_2	3.72 ± 0.09^{Ac}	3.68 ± 0.06^{Abc}	3.64 ± 0.11^{Ab}	2.59 ± 0.01^{Aab}	2.57 ± 0.06^{Aab}	2.52 ± 0.05^{Aa}	2.49 ± 0.04^{Aa}

Treatments		Mean number of bacteria in log_{10} cfu g^{-1} gut tissue						
		0th day	15th day	30th day	45th day	60th day	75th day	90th day
	D_3	3.67 ± 0.04^{Ad}	3.63 ± 0.08^{Ad}	3.58 ± 0.09^{Ac}	2.88 ± 0.01^{Bb}	2.82 ± 0.01^{Bab}	2.78 ± 0.01^{Cab}	2.41 ± 0.06^{Aa}
	D_4	3.74 ± 0.05^{Ac}	3.72 ± 0.08^{Ac}	3.69 ± 0.04^{Abc}	2.81 ± 0.07^{Bb}	2.79 ± 0.02^{Ba}	2.72 ± 0.08^{Ca}	2.69 ± 0.06^{Ba}
	D_5	3.71 ± 0.05^{Ac}	3.68 ± 0.01^{Ac}	3.61 ± 0.03^{Abc}	2.74 ± 0.05^{Bb}	2.71 ± 0.08^{Bab}	2.63 ± 0.05^{Ba}	2.58 ± 0.08^{Aba}
PPC	D_0	3.96 ± 0.01^{Ab}	3.93 ± 0.11^{Bb}	3.86 ± 0.07^{Bab}	3.79 ± 0.09^{Ca}	3.72 ± 0.09^{Ca}	3.81 ± 0.11^{Cab}	3.84 ± 0.12^{Bab}
(Presumptive	D_1	3.94 ± 0.01^{Ad}	3.83 ± 0.08^{Abc}	3.78 ± 0.03^{ABc}	2.93 ± 0.06^{Bb}	2.86 ± 0.08^{Bb}	2.79 ± 0.05^{Bab}	2.29 ± 0.04^{Aa}
Psedumonas	D_2	3.92 ± 0.05^{Ac}	3.89 ± 0.05^{ABc}	3.83 ± 0.03^{Bbc}	2.69 ± 0.03^{Ab}	2.67 ± 0.01^{Aab}	2.67 ± 0.06^{Aab}	2.41 ± 0.03^{ABa}
Count)	D_3	3.89 ± 0.08^{Ac}	3.87 ± 0.09^{ABc}	3.76 ± 0.01^{Abc}	2.88 ± 0.01^{Bb}	2.82 ± 0.07^{Bb}	2.77 ± 0.08^{Bab}	2.2 ± 0.09^{Aa}
	D_4	3.94 ± 0.06^{Ad}	3.88 ± 0.04^{ABc}	3.79 ± 0.04^{ABb}	2.74 ± 0.01^{ABab}	2.72 ± 0.08^{ABab}	2.69 ± 0.02^{Aab}	2.42 ± 0.04^{ABa}
	D_5	3.96 ± 0.05^{Ad}	3.82 ± 0.03^{Ac}	3.72 ± 0.08^{Abc}	2.82 ± 0.07^{Bb}	2.79 ± 0.02^{ABb}	2.72 ± 0.08^{ABb}	2.45 ± 0.08^{ABa}
TCC (Total Coliform Count)	D_0	3.22 ± 0.04^{Aa}	3.26 ± 0.07^{Aa}	3.28 ± 0.09^{Aa}	3.32 ± 0.02^{Ca}	3.34 ± 0.01^{Ca}	3.38 ± 0.09^{Ca}	3.41 ± 0.05^{Ca}
	D_1	3.27 ± 0.07^{Ac}	3.19 ± 0.05^{Ab}	3.12 ± 0.12^{Ab}	2.89 ± 0.01^{BCab}	2.82 ± 0.07^{BCab}	2.82 ± 0.04^{Bab}	2.76 ± 0.05^{Ba}
	D_2	3.24 ± 0.03^{Ab}	3.2 ± 0.01^{Ab}	3.18 ± 0.1^{Ab}	2.65 ± 0.02^{Bab}	2.61 ± 0.06^{Ba}	2.59 ± 0.01^{Ba}	2.55 ± 0.04^{Ba}
	D_3	3.31 ± 0.02^{Ac}	3.16 ± 0.05^{Ab}	3.11 ± 0.02^{Ab}	2.54 ± 0.08^{ABab}	2.52 ± 0.04^{Bab}	2.49 ± 0.03^{ABa}	2.1 ± 0.03^{Aa}
	D_4	3.29 ± 0.08^{Ac}	3.18 ± 0.04^{Ab}	3.16 ± 0.02^{Ab}	2.26 ± 0.03^{Aab}	2.21 ± 0.01^{Aab}	2.17 ± 0.05^{Aa}	2.09 ± 0.08^{Aa}
	D_5	3.27 ± 0.03^{Ac}	3.22 ± 0.04^{Ab}	3.21 ± 0.04^{Ab}	2.42 ± 0.04^{ABa}	2.39 ± 0.05^{ABa}	2.37 ± 0.09^{ABa}	2.31 ± 0.01^{ABa}

Values (Mean ± SD) with different upper case superscripts denote significant difference among the treatments ($P<0.05$), and values (Mean ± SD) with different lower case superscripts denote significant difference during experimental period ($P<0.05$)

Table 12: Log counts per gram of probiotic strain *Bacillus licheniformis fb 11* in the supplemented diets during storage at 4 °C (Mitra, 2015b)

Treatments	Days of storage						
	0	15	30	45	60	75	90
D_1	$4.69±0.05^{Ae}$	$4.56±0.03^{Ae}$	$4.47±0.01^{ABe}$	$4.39±0.01^{ABe}$	$4.26±0.04^{Be}$	$4.07±0.06^{Be}$	$3.72±0.05^{Ce}$
D_2	$5.86±0.05^{Ad}$	$5.79±0.005^{Ad}$	$5.68±0.17^{ABd}$	$5.51±0.034^{ABd}$	$5.30±0.05^{Bd}$	$4.92±0.05^{Cd}$	$4.75±0.07^{Cd}$
D_3	$6.67±0.12^{Ac}$	$6.54±0.12^{Ac}$	$6.45±0.14^{ABc}$	$6.38±0.18^{ABc}$	$5.96±0.16^{Bc}$	$5.81±0.13^{BCc}$	$5.74±0.15^{Cc}$
D_4	$7.78±0.16^{Ab}$	$7.59±0.16^{Ab}$	$7.46±0.11^{ABb}$	$7.32±0.16^{ABb}$	$7.14±0.13^{Bb}$	$6.92±0.12^{Cb}$	$6.8±0.0.19^{Cb}$
D_5	$8.85±0.1^{Aa}$	$8.67±0.57^{Aa}$	$8.52±0.46^{Aa}$	$8.43±0.19^{ABa}$	$8.25±0.17^{Ba}$	$7.94±0.14^{BCa}$	$7.73±0.16^{Ca}$

Values (Mean ± SD, n = 3) with different upper case superscripts denote significant difference during the storage period at 4°C (P<0.05), and values (Mean ± SD, n = 3) with different lower case superscripts denote significant difference among the treatments (P<0.05)

Table 13: Water quality parameters (Mitra, 2015b)

Variables	Treatments					
	D_0	D_1	D_2	D_3	D_4	D_5
Temperature (°C)	27.3±0.5	27.8±0.25	27.4±0.65	27.6±0.5	27.3±0.53	27.9±0.5
pH	7.4±0.4	7.4±0.5	7.5±0.1	7.5±0.2	7.5±0.5	7.4±0.5
Ammonia (mgL^{-1})	0.073±0.0012	0.076±0.0011	0.07±0.0011	0.068±0.0013	0.072±0.0011	0.071±0.0011
Dissolved oxygen ($mg\ L^{-1}$)	6.6±0.3	6.8±0.5	6.4±0.5	6.9±0.6	6.4±0.2	6.7±0.5
Free Carbon dioxide ($mg\ L^{-1}$)	2.3±0.15	2.5±0.18	2.6±0.12	2.3±0.15	2.6±0.12	2.6±0.14
Total alkalinity ($mg\ L^{-1}$)	90±0.78	93±0.53	92±0.65	92±0.6	93±0.43	92±0.3
Total hardness ($mg\ L^{-1}$)	112 ±1.58	114±1.61	112±1.56	114±1.58	112±1.65	112±1.63

Values (Mean ±SD)

6

Reproductive Biology

The knowledge on the reproductive biology of fish is very important for productive aquaculture and scientific based fishery management. Manipulation of a fish's reproductive system under culture or captive conditions requires an understanding of natural spawning patterns and other influential factors. Reproductive biology i.e. fecundity, spawning, sex ratio etc. are among the important aspects of the biology of fishes, which must be understood to explain the variations in the level of populations as well as to make efforts to increase the amount of harvest (Azadi and Mamun, 2004). Considering the importance, we have discussed on the different aspects of reproductive biology of this species here.

6.1 Age at Maturity

The female *Chitala* attains maturity in the third year of life, male sometimes at the later part of the second year Chondar (1999). Sarkar et al. (2006a,b) also reported that maturity can be observed at a minimum size of 600 mm for male and 700 mm female. But the age structure and growth pattern of natural population of *C. chitala* was well documented by Sarkar et al. (2008) from different river basins of India i.e. Samaspur Bird Sanctuary (SBS, 25^0 97′N, 81^0 67′E), Katraniaghat Wildlife Sanctuary (KWS, 25^0 97′N, 25^0 97′E), river Ghagra (26^0 01′N, 83^0 22′E), Gomti (25^0 90′N 82^0 56′E), Bhagirathi(24^0 05′N, 88^0 06′E), Koshi (25^0 82′N, 81^0 26′E), Saryu (26^0 12′N, 82^0 45′E), Satluj (31^0 09′N, 74^0 56′E), Ganga (24^0 53′N, 88^0 10′E), Yamuna (25^0 41′N, 81^0 91′E). To describe the age structure total radii and number of the annuli of the scales below the dorsal fin and above the lateral line were studied. The seasonal changes of gonadal development the following categories was used according to Jhingran (1991) i.e. Stage I – virgin, Stage II – developing, Stage III – maturing-I,

Stage IV – maturing- II, Stage V – gravid, Stage VI – spawning ,Stage VII – spent. From the study, a maximum 6+ ages (1033.36–1073.63 mm) were recorded from different population. They also showed that the size at maturity and maturity percentage varied in males and females across different river basins. The male C. *chitala* attained maturity a year earlier than females i.e. at 2+ age (473.1–763.76 mm) whereas females matured at age 3+ (621.33-855.74 mm. length). No maturity was recorded in 1+ age (338.5-627.27 mm. length) in case of males and 1+ to 2+ (473.1-763.76 mm. length) age classes in females. The sudden increase of growth at the age 3+ in the samples of river Bhagirathi, may be due to the exposure of the fishes to favorable environmental conditions after 2+ age. But due the environmental degradation no individuals were observed more than 3+ ages from rest of the locations.

6.2 Sexual Dimorphism

Chonder (1999) described *Chitala* as bisexual. But Sarkar et al. (2006 b) reported that the male and female were not easily distinguishable. They observed that in most of the cases females were bigger in length than male. The dimorphic character of male and female *Chitala* is being summarized after Chonder (1999) and Sarkar et al. (2006 b).The male *Chitala* during the breeding season shows diffused but bright colour at the base of paired and anal fins. Ground colour of body varies from silvery white to steel grey depending upon habitats. Papilla is thin, muscular hard, conical in shape. It is more pointed and tipped with reddish colour. The males show diffused vent and red coloration at the base of paired and anal fins. Single testis is unilaterally placed on the left side of the body. But the female does not show any marked colouration at the base of paired and anal fins. Abdomen bulged externally and could be seen prominently disposed in compared to thus in male. Papilla was stouter, fleshy, thin walled and broad. It is less pointed and may not be tipped with red colour. The shape of the gonad was sac like structure in which the eggs were embedded like fimbrae. Fully mature female shows freely oozing ova. Jahan et al. (2015) also identified the matured male fish by a slightly pointed genital papilla, and females by a swollen abdomen and a reddish swollen vent. In addition, the maturity of the female can be confirmed by slightly pressing on the ventral side of the fish for oozing of eggs.

6.3 Gonadal Maturity, Structure and Breeding Periodicity

The gonadal maturity of Chital appears in a cyclic manner once in a year. Jhingran (1991) reported seven stages of gonadal development in *Chitala* where the stage I was considered as virgin, stage II was developing gonads, stage III was designated as first stage of maturity, stage IV was described as maturing stage II, gravid stage was reported as stage V, spawning condition of gonads was designated as stage VI and spent gonads were described as stage VII. But later Chonder (1999) elaborately described the condition of gonad and the timing of maturity phase in Indian environment. Stage I gonads can be found during the month of November to March. The peak time is November to December. The characteristic of testis and ovary is as follows- thin, transparent, greyish testis is a small compact mass extended left dorso lateral side of the coelom and hardly distinguishable from ovary. Transparent, thin, compact mass of ovary is greyish in colour small in size at left side of the body cavity. Ova are minute, microscopic, spherical and remain in clusters. The developing stage II gonad can be seen from December to March and the peak season is from December to January. The testis become slightly thicker than before, greyish white in colour and compact. The appearance of ovary becomes translucent, thick glandular compact mass and the spherical ova increased in size. Stage III (Maturing-I) can be observed from January to April. The peak season is in January to February with a thicker, triangular shaped testis which become enlarge and compact mass pale cream in colour. The ovary becomes triangular, opaque, prominently granulated and enlarged further in length and girth. Ova can be distinguished by naked eye, creamy yellowish in colour due to appearance of minute yolk granules. The stage IV (Maturing II) can be differentiated by the thicker and compact creamy mass of testis, extending anteriorly 3/5 th of the body cavity. The ovary becomes granulated, pale yellowish in colour extending anteriorly 3/5 th or more of the coelom and become thicker. The ova are moderately large with different size and spherical to globular yellow in colour densely packed with small and large yolk granules. This stage can be observed from February to May. The peak time is in March to April. The matured stage V can be seen during March to June and peak in mid April to mid May. The size of the testis become further increased with a soft firm texture and creamy in colour. Milt appears on slightly pressure on testis over abdomen. The ovary become further increased in size covering the body. The outline of the ovary is lobulated and prominently granulated. The opaque to transparent yellow ova are different in size and mixed together, round in shape. The spawning stage VI occurred during May to August. Peak time is in mid May to mid July. The rosy colour testis becomes extended fully with a firm, soft texture rosy. The oozing of milt can be seen by pressing the abdomen. The general appearance of the ovary is almost same as that of stage V but slightly softer and yellowish to pale brown in colour. Each ovum is of different size. The smaller one being yellowish, the larger one is transparent separated from each other and ready to shed. During May to August stage VII (Spent) can be observed. The peak time is in August, when the creamy white testis become shrunken, flabby and being empty. The ovary also becomes shrunken, flabby, reduced in size and pale yellow in colour. A few left over yellowish ova can appear.

Rahman (1989) also reported that chital breed in June and having an egg size of 3.0 to 4.5 mm in diameter. But Radheyshyam and Sarangi (2005) observed the diameter of fertilized eggs ranges were 4.8-8.2 mm. Hossain et al. (2006) cultured chital in ponds and found the fertilized eggs having average diameter of 4.5 mm. Singh et al. (1980) first identified three categories of ovarian eggs, namely immature ova which were 0.5-1.9 mm. in diameter, but mostly ranged between 0.5-1.5 mm. and they were 66%.maturing ova of about 7% with the diameter of 2.0-2.7 mm. and mature ova of 27% with the diameter ranged between 2.8-4.8 mm. But the ripe ova was globular, yellow 5 mm. in diameter with the glutinous surface projections to attach with hard substratum. Kohinoor et al. (2012) studied the histology of female gonad and ova diameter of *Chitala*. They found that the ovary was oval shaped and single lobed lying in the body cavity chital. The immature ovary was compact. But the matured one was filament like structure. But the ovary becomes matured in June showing filament like structure and yellowish in colour. The ova were categorised on the basis of size of diameter i.e. category I (0.04-0.168 mm) and category-II (1.00-4.00 mm). The smallest diameter of ovum was recorded 0.04 mm and the largest was 4.00 mm during the study period. During January to March, only category-I oocytes were found. Mean ova diameter was 0.117±0.02 mm and 0.120± 0.03 mm in February and March. The category II ovum was found in April with a rapid increase in diameter in May. It was observed that the abundance of category II oocytes were highest in June (34.21±6.54%) and lowest in April (8.86±3.38%). The mean ova diameter of Category I and II, were 0.122±0.02 mm and 2.51±0.54; 0.124±0.021 and 3.16±0.89 mm; 0.129±0.02 mm and 3.21±0.47 mm in the month of April, May and June, respectively. Based on the size and appearance of nucleolus and cytoplasm different developmental stages of oocyte i.e. oogonia (OG), early perinucleolus stage (EPN), late perinucleolus stage (LPN), cortical alveoli stage (CA), vitellogenic stage (VG) were observed. At the earlier stages of the study period (January/February), all oocytes were found in oogonial stage. Later on, the percentage of oogonia showed a decreasing trend in the successive months. The highest percentage early perinucleolus stage (EPN) was found in May (31.00±2.65 %) with a decreasing trend in the following months. The percentage of late perinucleolus (LPN) stage was highest in April (7.67±1.53%) and then their appearance was found to decrease gradually. But the early and late perinucleolus oogonia stage showed a decreasing trend from February to June. Cortical alveoli stage (CA) appeared from April and turned greatest in May (10.67±2.08%). Vitellogenic stage (VG) appeared in the month of May and chronologically increased up to June (33.67±3.51%). Atretic oocytes (AO) were not reported during the study period.

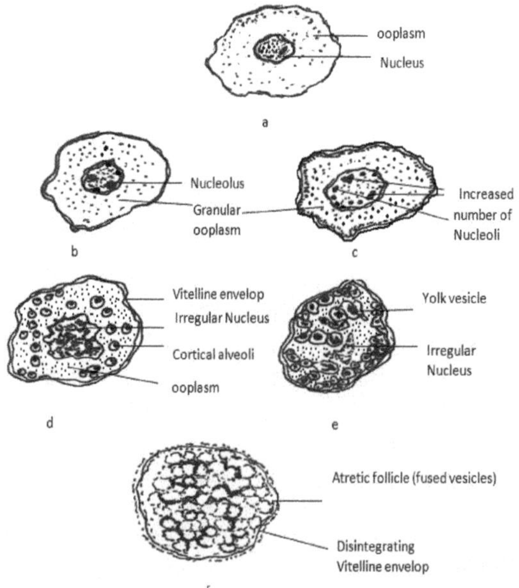

Fig18: General characteristics of different stages during oocyte development in fish (a) oogonia (b) early perinucleolus stage with increase number and size of nucleolus around periphery (c) late perinucleolus stage with further increase in number of nucleoli (d) cortical alveolus stage with formation of the vitelline envelop, irregular nucleus and proliferated alveoli (e) vitellogenic stage with yolk vesicles, irregular degenerating nucleus and prominent vitelline membrane (f) atretic stage with disintegrating vitelline membrane and fused vesicles. © Anisa Mitra

6.4 Gonadosomatic Index and Fecundity

The estimation of gonadosomatic index and fecundity of a fish is essential for evaluating the commercial potentialities of its stock, life history, practical culture and actual management of the fishery (Rahimibashar et al., 2012). Gonadosomatic index (GSI) is one of the important parameters of the fish biology, which gives the detail idea regarding the fish reproduction, reproductive status of the species and help in determining breeding period of fish (Shankar and Kulkarni, 2005). Gonads undergoing regular seasonal cyclic changes in weight, particularly in females indicate the spawning season (Dadzie et al., 2000). In *Chitala* the average gonado somatic indices of male and female are about 2.8% and 5% respectively in fully matured fish. Singh et al. (1980) reported the weight of ovary of a 8 kg chital as 400 g only and GSI 1:7.5. Hossain et al. (2006) observed the fecundity of *Chitala* as 1.371 during their breeding and fry nursing experiment. They have also correlated this low fecundity to the bigger size of eggs, single lobed ovary and nest building behaviour and parental care of *Chitala*. Kohinoor et al. (2012) studied the GSI and fecundity for a period of six months from January to June, 2010. The mean GSI values for female chital were found to range between 0.20±0.013 and 4.63±0.50. The highest GSI value was found in June indicating its peak period of ovarian growth. The fecundity was ranged from 8,238 to 18,569 (mean 13,052±4607) with body weight range from 1,296 to 2,360 (mean 1,742.50±474.44 g) while the relative fecundity was 5.65 to 14.33.

6.5 Sex Ratio

The studies on sex ratio provides information on the proportion of male to female fish in a population, it also indicates the dominance of sex in a given population and the basic information is important for studying fish reproductive potential and stock size assessment (Vicentini and Araujo, 2003). Azadi et al. (1995) described the sex ratio of females to males in *Chitala* as 0.721:1.0. Whereas Radhyasyam and Sarangi (2005), Sarkar et al. (2006b) and Jahan et al. (2015) observed the sex ratio as 2:1 for male: female during performing the captive breeding experiment.

6.6 Breeding

Southwell and Prasad (1918) described the breeding habits of Chital in the river Ganga. According to them during egg shedding in the river female lies close proximity to the objects like stones, brick walls, tin carriers, wooden box or any other hard objects lying submerged at the corner/sides of the pond on which the glutinous ova are to be deposited, first the body of the fish being inclined at a certain angle to the vertical. It has been also reported that the spawning appears to be prolific inside tin carrier and stone blocks which provide better environment and protection to the eggs as nest. No deposition of eggs is reported on aquatic weeds. The male later on releases milt over the deposited eggs. 300-500 eggs are usually laid at a time and nest is very carefully protected by the parent fishes. A maximum of 100 eggs/25 cm^2 surface of the substratum could be obtained among which 60% of the eggs generally fertilized. Chital breeds also in lentic environment. In pond its spawning commences immediately after the early monsoon rains following dilution of pond water to some extent. Singh et al. (1980) reported the spawning of Chital in ponds at Cuttack (Orissa) in June-July after early rains when the pond water was slightly diluted. But there is no report whether chital performs any spawning migration in shoal or in pair in the rivers, reservoirs and in lotic environments. However, in captivity such as ponds, tanks etc the spawning pairs move together near the spawning grounds in search of suitable nests the hard objects. (Chonder, 1999). Radhashyam and Sarangi (2005) investigated to refine the technique of quality seed production of Chital through controlled breeding. They carried out the experiment in a private fish farm in Orissa. The riverine stock was reared in pond till attaining maturity and transferred to a carp brood fish pond for intensive management. The brood fishes were fed once daily with a mixture of groundnut oil cake and rice bran @ 3% of body weight. Water was exchanged 10-15% fortnightly from ground water source. The female-male ratio was kept at 1:2.The female brood fish was injected with WOVA-FH fish hormone @ 0.3 ml kg^{-1} body weight and males @ 0.15 ml kg^{-1}. Fertilized eggs were collected through bamboo poles fixed vertically in a well prepared pond of 0.08 ha (130 cm deep) by pushing about 25 cm deep into pond sediment. The female fish bred naturally after a gap of 12 days of first breeding. Hatching was carried out in a mini circular incubation pool. 60.4-67.5% eggs were fertilized and hatchling were harversted. Another breeding experiment was carried out in captivity by Sarkar et al. (2006b) in which wild *C. chitala* were collected from natural habitats and induced to spawn under captivity by injecting three different doses (1.5, 1.0 and 0.5 ml kg^{-1} body weight) of synthetic hormone Ovaprim intramuscularly. Artificial breeding pool was prepared for each set by encircling area (20 × 5 m) with mosquito net, where wooden country boat (8 × 4 × 2.5 feet) with surface area 48.5 sq. feet was placed inside the breeding pool. Discrete spawning behavior was noticed in the experimental sets with different hormonal dose but no spawning activity was noticed in case of the control set. The physico-chemical parameters of broodstock pond were noted as the following: air temperature (30 ± 1.1^0 C), water temperature (31 ± 2.21^0 C), pH 7.5 ± 0.23, dissolved oxygen (8.0 ± 2.3 ppm), free CO_2 (2.3 ± 0.5 ppm) and turbidity (2.5 ± 11 cm). Different physico-chemical parameters of breeding pool were recorded as; air temperature 30 ± 1.0^{0}C, water temperature 29.0 ± 2.2^0 C, pH 7.5 ± 0.2, dissolved oxygen 8.0 ± 1.3 ppm, free CO_2 2.3 ± 0.5 ppm, turbidity 5.0 ± 1.1 cm, alkalinity 60.0 ± 5.0 ppm and water hardness 200 ± 10.0 ppm. The fertilization rate varied from 48.86–80.2% and total number of spawned eggs in two set of experiments were estimated to be 81,034. The average number of eggs deposited 15 ± 2.1/square inches. The fertilized eggs were large in size (4.5 ± 0.05 mm), adhesive and attached to the hard substratum. The eggs hatched out between 168–192 h after fertilization and about 33,639 hatchlings were produced. The percentage survival of hatchlings varied from 42.2 to 65.60.

Hossain et al. (2006) described captive breeding becomes most effective, when cemented tank was used instead of bamboo pole, plastic barrel and barrel made of tin for collection of the fertilized eggs, increase fertilization and hatching percentage.

Another controlled breeding experiment of *C. chitala* was conducted by Jahan et al. (2015) in the breeding field complex under captive condition. During early

March, After preparing the experimental pond GIFT tilapia broods had been stocked @ 10,000 ha^{-1} at sex ratio 2:1 (female: male) with SIS (Small Indigenous Species) of fish 25 $kgha^{-1}$ which acted as the natural food to *C. chitala*. Brood tilapia and SIS were fed with supplementary feed containing 30% crude protein. Water was exchanged 10-15% fortnightly. The water quality parameters of the brood pond was as follows: pH 6.6-7.5, dissolved oxygen 5-6 mgL^{-1} and temperature 28 C-31^{0}C. In April, sexually matured brood fishes of *C. chitala* (2.5 kg to 4kg) were then released into the earthen breeding pond (0.08 ha) at a sex ratio 1:2 (female: male). Breeding pool was prepared during the third week of July, where wooden boards (27"×7"×1.5") were used as egg collector device. Approximately 48 to 72 h after breeding pool preparation, the substrates were checked for their ovulatory response. After fertilization, the swollen eggs were found to be attached readily to substrates. The fertilized eggs were then immediately transferred to three cemented tank containing glass nylon hapa with continuous water supply throughout the incubation phase. The unfertilized eggs were opaque, spherical and whitish in colour measuring 2.0 to 2.5 mm in diameter. While the fertilized eggs were cream color, transparent, spherical, and adhesive in nature. Immediately after fertilization the diameter of the egg increased owing to slight swelling of the egg which ranged between 4.2 to 5.0 mm.

6.7 Parental Care

Parental care is the investment in young after fertilization, in which parent provides the care to their young and this also varies greatly from fish species to fish species. The impressive variation in parental-care tactics has made fishes an excellent group for testing our understanding of how parental care evolves. The benefit of parental care is that it improves survival and development of young. In case of *Chitala* both the male and female take active part in parental care and it is very prominent. The guarding parent(s) sometimes become furious and even attack their enemies (Chonder, 1999). Radhesyam and Sarangi (2005) described an intensive parental care as a significant part of breeding activity in Chital. They observed that after spawning Chital floats around the breeding ground and if the egg collectors removed from the breeding ground the fish continued to hover the breeding site exhibiting surfacing rolling over and exposing their broad silvery side as a clear sign of parental care. Parental care by male fish was observed by Sarkar et al. (2006) throughout the hatching period by fanning over the eggs, keeping it aerated and protected during performing the captive breeding experiment. Hossain et al. (2006) also reported that fertilized eggs hatch out by the fanning activity of the parental fish which helps to supply dissolved oxygen. But it is not known whether the parental care continues to be on their fry (Southwell and Prasad, 1918). According to Balshine and Sloman (2011) biparental care is thought to arise from male-only care when females lay very large eggs and the benefits of defense by two parents outweighs the female fecundity costs of care. Biparental care is argued to lead to female-only care when male future mating opportunities increase. In *Chitala* as the females lay large sized eggs (3.0-4.5 mm diameter), & fecundity is low probably its biparental care may compensate its fecundity cost & increase mating chance.

7

Developmental Biology

The detailed information on larval development during early life history stages of the new candidate species helps to improve the production of high quality juveniles and adults (Çoban et al., 2009). For a better understanding on the development of *Chitala* in this section we have gathered all the available information on developmental phases of this species. The first report on the developmental progress in *Chitala* was documented by Southwell and Prasad (1919). They also showed that the embryonic development and hatching of eggs are slow in this fish species and the hatching process is prior in natural condition at 30-33^0 C than the laboratory condition. Southwell and Prasad (1919) and Singh et al. (1980) well documented the embryonic and larval development process elaborately. They have described that the fully matured embryo (5-5.2 mm. diameter) had surface projections. But in advanced stage the spherical large yolk became oblong in shape. Completely developed embryo showed very sluggish movement with a large yolk sac. The embryo of 7.1 mm. diameter showed differentiated head, positioned over ventral yolk sac, with a continuous embryonic fin fold. The ear and heart was formed. A small opening of mouth and straight tube like gut appeared. The air sac was oval shaped and rudimentary branchial arches were observed. Myotomes were 57 in number at this stage. At the time of hatching head and the body with large yolk material showed a rapid movement of tail to hatch out. The newly hatched larva (total length 13.8 mm., head length 2.1 mm., body depth 1.6 mm., length of air bladder 1.2 mm.) had a narrow and elongated body with broad head and well developed brain. Embryonic fin fold increased in size, caudal fin showed rudimentary fin folds. Gill slits were well developed. Small opening of mouth was appeared at the ventral position. The Stage 1 larva (total length 14.2 mm., head length 2.5 mm., body depth 1.8 mm., length of air bladder 1.2 mm.) showed separated head from reduced yolk sac. Body was opaque with minute pigmentation. Mouth shifted to lateral position. The impressions of jaws were visible. The eyes were partly closed. The appearence of fin rays at the future anal fin can be seen. The gill filaments were with gill rackers. The stage 2 larva (total length 17.1 mm., head length 3.0 mm., body depth 2.6 mm., length of air bladder 1.6 mm.) had a reduced yolk sac with a dorsal fin made up of rudimentary fin rays. The size of the pectoral (1.5 mm), anal (11mm.) and caudal fin (1.2mm) increased in length. Larva-stage 3 (total length 18.4 mm., head length 3.3 mm., body depth 3.2 mm., length of air bladder 2 mm.) had gill filaments on all the four gill arches. The jaws were almost formed bearing the teeth. The operculum markedly separated on the head. Anal and caudal fin rays were well developed. Fin rays on the dorsal fin developed to the median fin fold. Dorsal fin (0.8 mm.) was not fully developed. But the pectoral (1.6 mm.), anal (11.3 mm.) and caudal fin (1.4 mm.) were developed. Better development of alimentary tract, liver and air bladder can be seen during this stage. During the post larval stage (total length 19.5 mm., head length 4.5 mm., body depth 4.1 mm., length of air bladder 4.3 mm.) yolk sac was completely absorbed. Dorsal, pectoral, anal and caudal fin were developed further with definite fin rays. Operculum was also well developed. Mouth well developed with jaw bones and teeth on maxillaries, dentaries, vomers and palatines. Heart chambers become prominent. Alimentary tract differentiated into stomach, pyloric caeca and associated with well-formed liver. Kidneys become distinguished. But the genitals were not prominent.

Sarkar et al. (2009a) have also well described the morphological development of early and late life stages of *Chitala* from the captive bred and riverine population of Bhagirathi, West Bengal. They reported that the fertilized eggs (4.01 to 5.05 mm in diameter) were spherical, adhesive and milky pale yellow in colour, transparent in appearance. After 50 h of fertilization first cleavage was observed with two uniform blastomeres were formed. The second cleavage appeared four hours later, third cleavage was after 54 h and the fourth cleavage observed after 68 hrs of fertilization. The sixteen-celled stage was observed at about 75 h and thirty-two celled stage appeared after 80 h of fertilization. The dome shaped blastoderm gradually spreads over the yolk mass attaining the morula stage in about 85 h after fertilization. Blastoderm cells spread about one half of the yolk mass at 90 h and the yolk plug stages were noticeable after 95 h. The yolk sacs get compressed at about 100 h and the embryonic rudiments were noticeable approximately 105 h after fertilization. After about 110 h of fertilization, cephalic and caudal ends differentiated incompletely. At about 115 h the protruding cephalic region with rudiments of notochord and arteries were appeared. At about 120 h the old embryo showed slow twitching movement and just before hatching, comma shaped notochord can be seen. The newly hatched larva of 1-3 day (Total length 5.00 mm, Total weight 0.03 g) attached with hard substratum and possesses a yolk sac (4.05 mm) with conspicuous network of blood capillaries. At 5-6 day (Total length 12.5mm ± 1.9, Total weight 0.08g ± 0.05) the rudimentary head appeared on yolk sac (3.85mm). The eye diameter was 0.40 mm. The colour of body was dark pink and still attached on the hard surface. The mouth appeared, upper and lower jaw were clearly visible. Larvae fed on the mucous released by the parent fish during the parental care. On

7-8 day (Total length 12.34 mm ± 1.08, Total weight 0.092 g ± 0.007) the diameter of the yolk sac became 3.64 mm. The head length and eye diameter was 2.8 and 0.55 mm respectively. Mouth is prominent with clearly visible eyes and rudimentary notochord. The alimentary tract became thicker and convoluted. Head still remain attached with yolk sac. At 9-10 day (Total length 12.57 mm ± 0.97, Total weight 0.097 g ± 0.05): The head length and eye diameter became 2.73 and 0.75 mm respectively. The deep yellow yolk sac became reduced with 3.42 mm diameter. Head well developed, notochord increased and anal fin slightly visible. Larvae started to get detached from the substratum. The yolk sac was absorbed by most of the larvae between 12 -14 day (Total length 13.8 mm ± 1.14 and Total weight 0.0985 g ± 0.06). The light brown coloured larvae were with clearly visible notochord and dorsal, pectoral and anal fins. But the fin rays were rudimentary. The larva had four gill arches in the opercular cavity. Air bladder become developed. Mouth was terminal and eyes were shining. Lateral line appeared on the lateral side of the body with some dark pigmentation. Between 18 - 20 day (Total length 22.28 mm ± 1.06 and Total weight 0.099 g ± 0.08, Body Depth 4.79 mm ± 0.21) dorsal, pectoral and anal fins became well developed with rudimentary rays. Gills were also well developed. Yolk sac completely absorbed and they start to accept exogenous feed. Several dark and light bands were present throughout the body. On 25 - 30 day (Total length 22.68 mm ± 1.30 and Total weight 0.12 g ± 0.03, BD 4.44 mm ± 0.017) the pectoral, pelvic, dorsal and anal fins developed with 7 - 8, 4 - 5, 5 - 7, and 30 - 60 fin rays. Relatively prominent transverse bars can be seen along the dorsal ridge. Dorsal profile becomes more humped than the previous stage. The presence of 2 - 12 dark spots also appeared near the tail and along the anal fin base.

Acoording to Jahan et al. (2015) the development of *C. chitala* eggs are very slow to attain their embryonic and larval stages. At day 1, the yolk sacs get compressed and the blastodisc formed as an indication of successful fertilization. They also reported that as the egg shell of *C. chitala* was thick, the early cleavage stages of embryonic development were not clearly visible under microscope at day 2. But the embryonic rudiments were noticeable 3 day after fertilization. In the 4 day old embryo, the cephalic region became prominent and rudiments of notochord and the tubular heart was appeared underneath the head. The heart actively started to pump. At day 5, the brain and optic bud were noticeable. The yolk mass showed elongation and slow twitching movement. Before hatching, at day 6 the embryo started vigorous twisting movement inside the egg and continuously beat the egg shell by the caudal region especially around the middle part of the body. This movement gradually became vigorous and the egg capsules were weakened. The incubation period was from 6-7 d at water temperature range of 28-31 C. However, it took about 7 days for the whole brood of egg to hatch out. The newly (1-2 day old) hatched larvae (15.5±0.2 mm in length and 0.05g weight) were transparent, faintly reddish in color. The head and body of larvae were bent around the large yolk sac and attached to the substratum. The yolk sac was oval in shape and yellowish in color. The diameter of yolk sac was 4.85mm. A functional heart with blood circulation was noticed. The mouth was not developed but eye was clearly visible and notochord rudimentary. At 3-4 day (length 18.2 ± 0.2 mm, weight 0.08g ± 0.05) the diameter of yolk sac was 3.64 mm. Head was adhered with yolk sac. The heart became more distinct. Circulation of body fluid was seen around the notochord in addition to the yolk. The mouth appeared, upper and lower jaw were clearly visible. The pigmentation was dark in the anterior region and malanophores are scattered on the yolk sac. Three to four days after hatching, the caudal fin fold was noticed. In 5-6 day old larvae (length 22.7 ± 0.1mm and weight 0.09 g ± 0.02) the bulged yolk became gradually elongated and yolk sac became smaller (2.37 mm. in diameter). The alimentary canal became thicker and convoluted. Head well developed and notochord increased. A thin membranous fin fold surrounded the caudal region and extended up to yolk sac. At this stage larvae started to get detached from the hard substratum. The larvae were converged in a cluster and a few of them started swimming. On 7-8 day (length 25.8 mm ± 1.09 and weight 0.10 g ± 0.05) the body colour of the larvae became light brown with clearly visible notochord with some dark pigmentation. The yolk sac was absorbed by most of the larvae. All the fins like dorsal, pectoral and anal were clearly visible but rays were not cleared and ventral fins were rudimentary developed. Abdominal portion was more segmented than earlier stage. The digestive system became fully developed and the larvae searched for feeding.

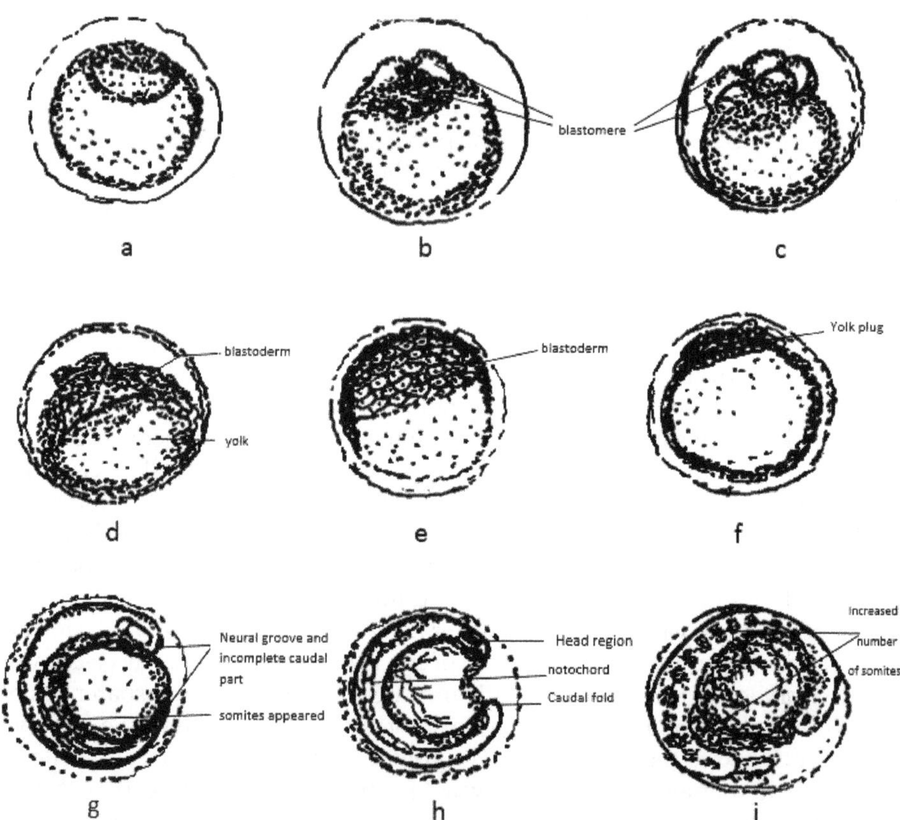

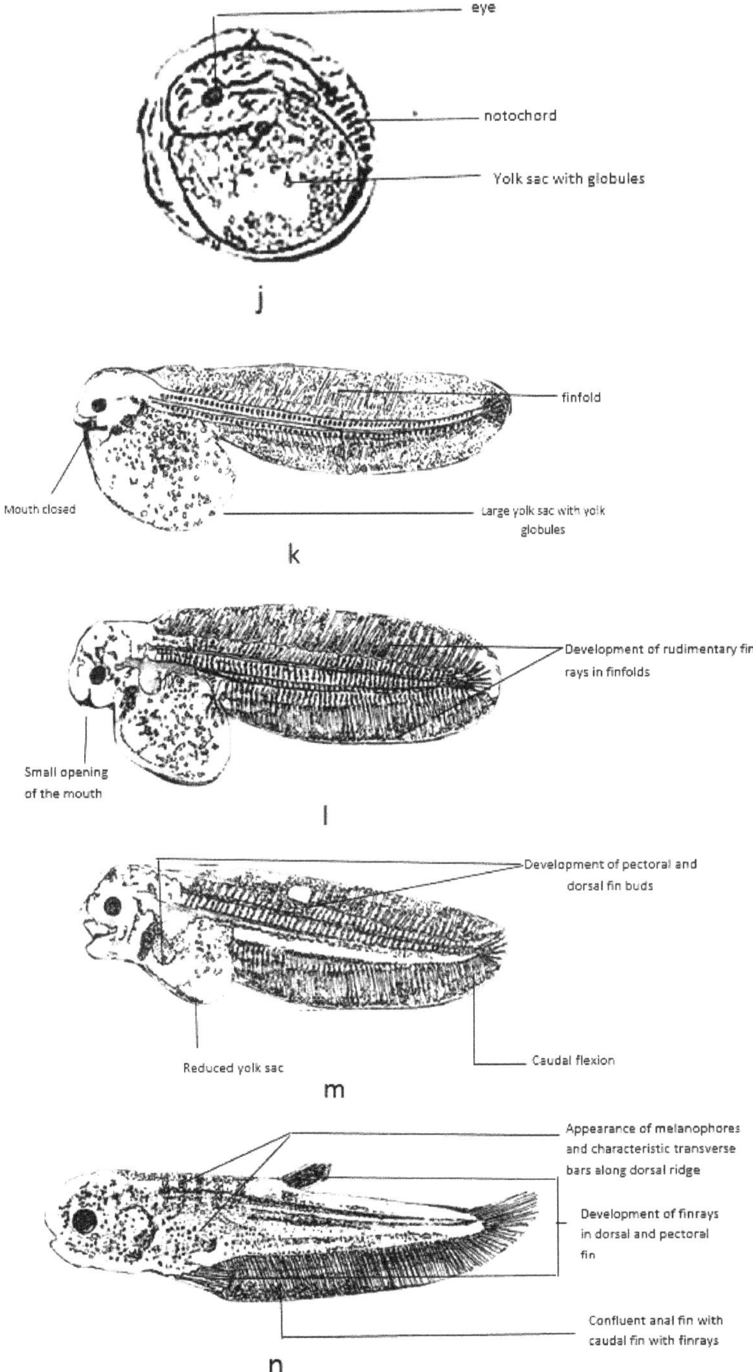
eye
notochord
Yolk sac with globules
j
finfold
Mouth closed
Large yolk sac with yolk globules
k
Development of rudimentary fin rays in finfolds
Small opening of the mouth
l
Development of pectoral and dorsal fin buds
Reduced yolk sac
Caudal flexion
m
Appearance of melanophores and characteristic transverse bars along dorsal ridge
Development of finrays in dorsal and pectoral fin
Confluent anal fin with caudal fin with finrays
n

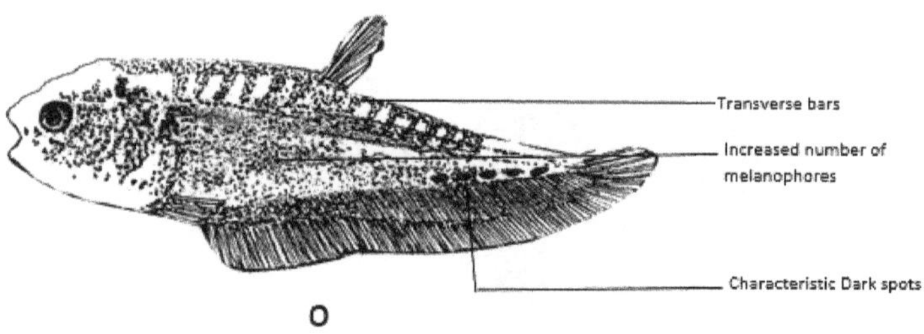

Fig 19: Developmental stages of *Chitala chitala.* (a-j) embryonic development: (a) fertilized egg (b) embryo with two blastomeres (50 h after fertilization) (c) morula stage (75 h after fertilization) (d) blastula stage (80 h after fertilization) (e) blastoderm spreads half of the yolk (90 h after fertilization) (f) yolk plug stage (95 h after fertilization) (g) embryonic rudiment; cephalic and caudal incomplete (110 h after fertilization) (h) protruding cephalic part with notochord and presence of blood vessels (115 h after fertilization) (i) embryo with further extended cephalic and caudal region (120 h after fertilization) (j) comma shaped embryo just before hatching). (k-p) larval and post larval stages: (k) newly hatched larva (1-3 DAH) (l) larva between 5-10 DAH (m) larva at 12 DAH (n) larva at 20 DAH (o) postlarva at 30 DAH (The diagrams were drawn based on the characters described above © Anisa Mitra)

8

Culture

Being a facultative air breather a large number of wild Chital can be produced in deep confined waters and can be adopted as a composite culture component even where carp culture may not be feasible (Chonder, 1999). Utilizing its piscivorous character Chital may be successfully used as a component in composite culture of carp. This species can be used to regulate the population imbalance that may be caused by wild breeding of common carp, abundance of other minnows and insects in ponds (Chaudhuri, 1975). Sarkar et al. (2006b) reported that a few attempts have been made to develop *Chitala* culture along with Indian major carps. But no scientific attempts were made to develop breeding protocols under captivity except some by fish farmers. Hence the cultural possibilities of this species in scientific lines are yet to be fully developed. The predatory habit of chital may not cause any danger to carp fingerlings in polyculture under judicial management and thus juveniles of chital can be incorporated in a suitable proportion in composite culture as a means of diversification of freshwater fish farming. Although effort to hatchery culture found in North and South 24 parganas, Burdwan, Murshidabad, Malda and Birbhum with the help of induced hypophysation (Mitra et al., 2012). The fry stocking for the culture of Chital is same as carp but it also depends on the seed availability. Cage culture of this species is feasible as per the guideline of the National Fisheries development board (2016) though few attempt has been made till date.

Few questionnaires were asked to the farmers and the fishery officials and their answers to understand present culture status in India.

Wild stock hot spot of *Chitala* in India and culture sites in west bengal	Uttar Pradesh, North Bengal, lower Assam, Tripura are traditional hot-spot for wild stock. Malda (Gangadharpur, Manickchak, Habibpur) uttar and dakshin dinajpur, Jalpaiguri, Coohbehar Burdwan (Purbasthali), Murshidabad (Farraka,Dhulian), are prime areas. Also available in few areas of Nadia and North 24 parganas. Effort to culture found in North and South 24 parganas, Burdwan, Murshidabad, Malda Birbhum
Culture type	Culture predominantly of extensive or semi-intensive type with feed, preferably live feed composite culture with carps are also done.
Demand in market-is it increasing?	Demand is sky rocketing, especially in cities and big hotels.
Price and marketable size	Marketable size> above 1.5 kg Preferred size> above 2 kg Rs. 750-900/ kg. But the taste varies with size.
% of wild population	About 70% of available in the market, comes from wild stock. Hence price is high and natural population declining. More than 50% population declined over 50 years
Investment in culture	Culture cost about 30000-35000 per bigha, excluding capital cost
Profitability	Profitability is high, depends on production and marketing. 10000-12000/- per bigha may be the profit margin
Export status	Export potential is good enough, but not yet explored properly. Export in national and Afro-Asian market is growing
% among total cultured fishes	Not measured properly, but national estimates reveals 2-7%
Kind of feed used during culture	Most preferred is live feed, sometimes small carp fry, which increases the cost. Prior to liberation of *chitala*, Tilapia and alike fast breeding species are liberated- in advance. They provide as live feed when *chitala* is released. But now a days some supplementary feeds are given which are also used for carps. So, there is a need to introduce feed in culture which is specific for this species understanding its requirement in different stages of life.

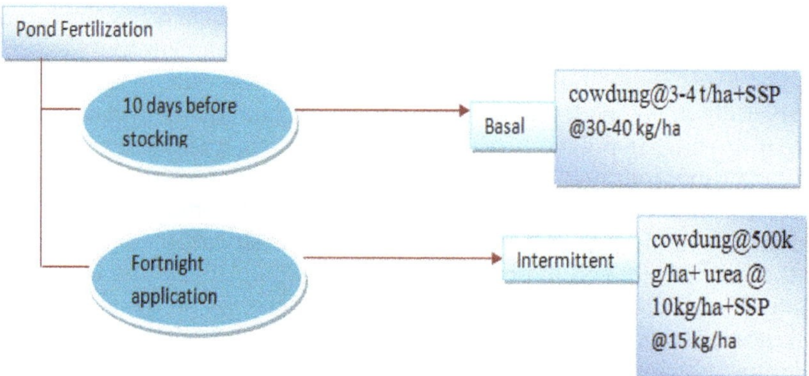

Fig 20: Schematic presentation of the pond preparation for *C. chitala* culture © Anisa Mitra

Fig 21: (a) Egg collection of *Chitala* from the pond (b) egg collected on brick (c) eggs on wooden boat surface (d) rearing in recirculatory system (e) rearing of *Chitala* in hapa (f) fingerling and (g) adult *Chitala* caught from fish farm (Photo credit Anisa Mitra and Pratap Kumar Mukhopadyay)

Epilogue

The current review work explicitly includes the available information on the morphological characters, food and feeding habit as well as reproductive biology and early developmental ontogeny of *Chitala*. But from the present study it is also evident that a very scanty information is available on the culture of this valuable species.

The study on growth pattern during early ontogeny of *Chitala* showed that most of the sensorial, feeding, respiratory and swimming systems developed rapidly during the first phase of development, supporting the hypothesis of energy and building materials allocation in priority to functional changes rather than size increase. The subsequent developmental period was mainly categorised by the flexion of notochord, the rapid and intense growth of the trunk and tail segments, the acquisition of adult axial muscle distribution, completion of gill filament development with a rapid development of fin rays, changes in body shape, locomotive ability and feeding techniques. All these changes in the growth trajectories of *C. chitala*, corroborated the transition from larvae to metamorphosing juvenile phenotype. This may act as a potential source of information to describe ontogenetic landmarks, which can be used to facilitate experimental research for developing the intensive culture of the species. Moreover, the analysis of growth patterns provides valuable information on the changes in functional demands throughout ontogeny, which could be useful for aquaculturists to design more suitable rearing conditions for particular life stages. Although, variations in the timing of ontogenetic events and structure formation can occur based on different environmental conditions (e.g. temperature and food availability), further research is needed to identify ossification processes and squamation patterns to define other important landmarks for the onset of the juvenile period more precisely.

The studies on feeding habit of different natural populations of *Chitala* showed that the species is primarily a carnivorous predator. But the capability of digesting diverse food components helps the species to widen the food spectrum in response to the habitat availability (Sarkar et al. 2009b). To optimize its present rearing protocol, different contemporary methodologies were used to study the development of gut histolomorphology, digestive enzyme profile, gut microflora and its role in digestion and health of fish. In *Chitala* around 12 DAH was found to be the best time to introduce first feeding. This is very important to establish the time of commencement of exogenous feeding to optimize the rearing strategy. But till date no reports are available on formulation of the age specific microparticulate diet for early life stages which is very important to reduce early life stage mortalities. Further researches can be conducted on the sex specific study of feeding and digestion especially on the nutritional requirement of gravid and spawning female fish to improve the quality of seed. Significant alterations in specific activities of major digestive enzymes were measured using biochemical techniques to develop a cost-effective formulated feed and feeding schedule to promote rapid and efficient growth rates under intensive farming. Amylase, lipase, alkaline protease activities were detected on 1 DAH but acid protease activity was detected on 12 DAH correlated with the mouth opening. The less efficient intracellular and alkaline digestion was substituted by a more efficient acidic digestion inside the gastric glands occurred at 12 DAH in *C. chitala*. Thus we can say that the digestive enzymes can act as comparative indicator during the development of *Chitala*. The pattern of digestive enzymatic activities in featherback larvae also reflected its ability to adapt with the diet during ontogenetic shift. This enables the species as a good candidate for aquaculture diversification using the compound diet from the early stage of development. Though detailed studies are necessary on the characterization of the digestive enzymes and histochemistry of gut development for better understanding of digestive capability in fishes.

Understanding the importance of Gut micro flora in the digestive process and in growth of the fishes, total eight types of bacterial flora were isolated from the *C. chitala* intestine namely *Bacillus* spp., *Vibrio* spp., *Aeromonas* spp., *Pseudomonas* spp., *Acinetobacter* spp., *Alcaligenes* spp., *Flavobacterium* spp., *Enterobacter* spp. This variation in the microflora of *C. chitala* might contribute to the incorporation of these bacteria in commercial aquaculture as supplement in formulated fish feed or in form of bacteria biofilm to achieve colonization in the fish gut at a higher degree. However, further researches involving these potent bacterial strains should be conducted for evaluating their efficacy under actual farm conditions. The formulated feed with *B. licheniformis fb 11* (strain isolated from *C. chitala* intestine) as probiotic feed additive at the concentration of $5X10^6$ CFU g^{-1} improved the digestive efficiency and growth performance significantly in *C. chitala* juvenile by modulating intestinal microflora without water quality degradation. Hence this may be also useful as an ecologically favourable alternative to live feeds in case of *C. chitala* juvenile. Further studies should be conducted on effects of Bacillus spp. as probiotic supplement in both live feed and/or in water for specific immune response, disease resistance and other biochemical parameter analysis.

Studies have also showed that the early life stages of *C. chitala* can be reared successfully in confinements like hapas and fibreglass reinforcement tanks by using fish eggs, dry tubifex, zooplanktons as a potential diet. But the reliability of the larval rearing protocols in recirculatory system is still at experimental level (Sarkar et al. 2008). Hence the species deserves further study on its suitability for commercial culture in confinement. Further investigations needed for optimization of stocking density, improving diet presentation, satiation level, feeding strategies, compatibility with other cultivable species during different life stages before adapting in to the culture system.

In contrast to the food and feeding habit relatively less information is available on different aspects of reproductive biology such as sex-ratio, breeding periodicity, gonadal development and histology, breeding behaviour and fecundity. Due to non-availability of seeds in natural waters and difficulty in artificial breeding of this fish, commercialization of this species has not been successfully achieved till date. In this aspect it is very important to conserve this endangered species in a sustainable manner. Several attempts have been made to develop its culture along with Indian major carps to develop breeding protocols under captivity. The studies showed that the fish was successfully bred under captivity and captive bred populations can be maintained. But the possibility of using synthetic fish hormone Ovaprim and WOVA-FH for effective induced spawning and seed production of *C. chitala* are still at experimental level. Evidently, ranching programme could be undertaken for species restoration and conservation. The successful development of protocols for captive breeding is likely to pave way towards commercialization of the technology, which may introduce an exciting entrepreneurial area (Sarkar et al., 2008). It has been also observed that several reports are available on the development of *Chitala*. But the differences in the timing of development may be due to the environmental conditions, nutritional status or for the stock difference of the brood fish. Hence further research is needed to determine the timing of different developmental stages more accurately and the key factors associated with it. Although the effect of different environmental variables i.e. rainfall, temperatures, water flow, pH on growth, reproduction and development of *Chitala* has not been also documented, which is very important for successful culture under captivity and subsequent conservation of any fish species information. No studies have been also performed till date regarding the disease and its management in *Chitala* which is very crucial to culture in captivity.

Chitala can tolerate wide range for temperature, dissolved oxygen and pH, which make this species as potential candidate for aquaculture. But the main challenge for its aquaculture is its slow growth rate in captive condition. Studies have been performed to promote its growth rate in early life stages and in fingerling but no success to date have been achieved in field condition. Thus, further study is needed to uphold its captive culture as only capture fishery may not be adequate to support its trade as a food fish in the long run.

All these information on different biological attributes of *C. chitala* would be useful for fishery biologists to execute adequate regulations for sustainable fishery management and conservation of this candidate species. Hence development of culture system for such an important species will open up new avenues for its sustainable production from inland waters. It is therefore imperative that aquaculturists, fishery entrepreneurs be little more proactive in this direction in utilizing the various productive fishery resources available in form of ponds, lakes, community tanks etc. with efficient use of nutrients and water to boost up diversified fish production. With this kind of approach it might be possible that a valuable species like *Chitala* can be rescued from being endangered. This will also enable in making available such an important food fish to the consumer providing nutritional security and livelihood support in the country.

References

Albrecht, M.P., Ferreira, M.F.N. and Caramaschi, E.P. 2001. Anatomical features and histology of the digestive tract of two related neotropical omnivorous fishes (Characiformes; Anostomidae). Journal of Fish Biology 58: 419-430.

Alikunhi, K. H. 1957. Fish Culture in India, Farmers Bulletin. Indian Council of Agricultural Research 20:144.

Anders, P.J. 1998. Conservation aquaculture and endangered species: Can objective science prevail over risk anxiety? Fisheries 23:28–31

Anson, M. L. 1938. The estimation of pepsin, trypsin, papain and cathepsin with hemoglobin. Journal of General Physiology 22: 79-89.

AOAC. 2000. Official Method of Analysis. 17th Edn., Association of Analytical Chemists,. AOAC, Washington, DC, USA.

APHA. 1992. Standard methods for the examination of water and waste water, 19th Edn. American Public Health Association, Washington, DC, USA.

APHA.2005. Standard methods for examination of water and wastewater, 21st Edn. American Public Health Association, Washington DC, USA.

Applebaum, S.L., Perez, R., Lazo, J.P. and Holt, G.J. 2001. Characterization of chymotrypsin activity during early ontogeny of larval red drum (*Sciaenops ocellatus*). Fish Physiology and Biochemistry 25:291-300.

Ayyappan, S., Raizada, S. and Reddy, A.K. 2001. Captive breeding and culture of new species of aquaculture. In: Captive Breeding for Aquaculture and Fish Germplasm Conservation (Eds. Ponniah, A.G., Lal, K.K. and Basheer, V.S.) pp. 1-20. National Bureau of Fish Genetic Resources (NBFGR), Lucknow, India.

Azadi, M.A., Mahamood, N. and Shafi, M. 1994.Studies on the age and growth of Chital, *Notopterus chitala* (Ham.) from the Kaptai Reservoir, Bangladesh. Chittagong University Studies 18 (2): 197-205.

Azadi M.A. and Mamun, A. 2004. Reproductive biology of the cyprinid, *Amblypharyngodon mola* (Hamilton) from the Kaptai Reservoir, Bangladesh. Pakistan Journal of Biological Sciences 7 (10): 1727-1729.

Baglole, C.J., Murray, H.M., Goff, G.P. and Wright, G.M. 1997. Ontogeny of the digestive tract during larval development of yellowtail flounder: a light microscopic and mucous histochemical study. Journal of Fish Biology 51: 1201-34.

Bairagi, A., Ghosh, K., Sen, S.K. and Ray, A.K. 2002. Enzyme producing bacterial flora isolated from fish digestive tracts. Aquaculture International 10: 109–121.

Balcázar, J. L., Blas, I., Ruiz-Zarzuela, I., Cunningham, D., Vendrell, D. and Múzquiz, J. L. 2006. The role of probiotics in aquaculture. Veternary Microbiology 114: 173–186.

Balshine, S. and Solman, K.A.1995. Parental care in fishes. Encyclopedia of Fish Physiology: From Genome to Environment 1: 670-677.

Basurco, B. and Abellan, E.1999. Finfish species diversification in the context of Mediterranean marine fish farming development. In: Cahier Options Méditerranennes 24: Marine Finfish Species diversification: Current Situation and Prospects in Mediterranean Aquaculture, C.I.H.E.A.M. (Eds. Abellan, E. and Basurco, B.) pp.9-25. Zaragoza, Spain.

Ben Khemis, I., Zouiten, D., Bebes, R., and Kamoun, F.2006. Larval rearing and weaning of thick lipped grey mullet (*Chelon labrosus*) in mesocosm with semi-extensive technology. Aquaculture 259: 190-201.

Bergy, D.H. 1986. Bergey's manual of systemic bacteriology. (Eds. Holt, J.H., Sneath, P.H. and Krieg, N.R.) Vol. 2 .Lipponcott Williams and Wilkins. 636pp.

Bernfeld, P. 1951. Amylases α and β. In: Methods in Enzymology. (Eds. Colowick, P. and Kaplan, N.O.) Vol.1. pp.149-157. Academic Press, New York.

Betti, P., Machinandiarena, L. and Ehrlich, M.D. 2009. Larval development of Argentine hake *Merluccius hubbsi.* Journal of Fish Biology 74: 235-249.

Beveridge, M.C.M. and Little, D.C. 2002. History of aquaculture in traditional societies. In: Ecological Aquaculture (Ed. Costa-Pierce, B.A.) pp. 3-29. Blackwell Science, Oxford.

Bhatt, J.P., Nautiyal, P. and Singh, H.R. 2000. Population structure of Himalayan mahseer, a large cyprinid fish in the regulated foothill section of the river Ganga. Fisheries Research, 44(3): 269.

Blaxter, J.H.S. 1988. Pattern and variety in development. In: Fish physiology. (Eds. Hoar, H. S. and Randall, D. J.) Vol. 11. pp. 1-58. Academic Press, New York.

Bone, Q., Marshall, N.B. and Blaxter, J.H.S. 1995. Locomotion. In: Biology of fishes. 2nd edn. pp. 44-78.Chapman and Hall, London.

Borlongan, I.G. 1990. Studies on the digestive lipases of milkfish, *Chanos chanos*, Aquaculture 89: 315–325.

Cahu, C.L. and Zambonino-Infante, J.L. 2001. Substitution of live food by formulated diet in marine fish larvae. Aquaculture 200: 161-180.

CAMP. 1998. Conservation Assessment and Management Plan Workshop Report. Zoo outreach organization /CBSG and National Bureau of Fish Genetic Resources Lucknow pp.1-158, Lucknow, India.

Cara, J.B., Moyano, F.J., Cardenas, S., Fernandez-Diaz, C. and Yufera, M. 2003. Assessment of digestive enzyme activities during larval development of white bream. Journal of Fish Biology 63: 48-58.

Celik, P. and Cirik, S. 2011. Allometric growth in serpae tetra (*Hyphessobrycon serpae*) larvae. Journal of Animal Veterinary Advances 10:2267-2270.

Chakrabarti, R., Rathore, R.M. and Kumar, S. 2006a. Study of digestive enzyme activities and partial characterization of digestive proteases in a freshwater teleost, *Labeo rohita*, during early ontogeny. Aquaculture Nutrition 12: 35–43.

Chakrabarti, R., Rathore, Mittal, P. and Kumar, S. 2006b. Functional changes in digestive enzymes and characterization of proteases of silver carp (♂) and bighead carp (♀) hybrid, during early ontogeny. Aquaculture 253: 694–702.

Chandy, M .1961. Origin, distribution, phylogeny and interrelationship of Ophicephalid fishes (Snake head murrels) of India. Proceeding of Indian Science Congress 42 (3): 304.

Chaudhuri, H. 1975.A new high in fish production in India with record 1975 yields by composite fish culture in freshwater ponds. Aquaculture 5(6): 343-355.

Chonder S.L. 1999. Biology of Finfish and Shellfish. SCSC Publishers, Howrah, India.514 pp.

Cobán, D., Kamaci, H.O., Suzer, C. and Saka, A. 2009. Allometric growth in hatchery-reared gilthead seabream. North American Journal of Aquaculture 71: 189-196.

Dadzie, S., Abou-Seedo, F. and Al-Qattan, E. 2000. The food and feeding habits of the silver pomfret, *Pampus argenteus* (Euphrasen), in Kuwait waters. Applied Ichthyology 16: 61-67.

Das, S.M. and Moitra, S.K.1963. Studies on the food and feeding habits of some freshwater fishes of India. IV. A review on the food and feeding habits, with general conclusions. Ichthyologica 11(1-2): 107-115.

Datta Munshi, J. S. and Srivastava, M. P.1988. Natural history of fishes and systematics of freshwater fishes of India. Narendra Publ. House, Delhi.: i-xviii, 1-403.

Day, F. 1878. The fishes of India. Publisher, London. p. 703, pl. cl xxxii, fig. 4.

Day, F. 1889. The fauna of British India, including Ceylon and Burma. vol.1. p. 548. Fishes, Taylor and Francis, London.

de Silva, C. 1974. Development of the respiratory system in herring and plaice larvae. In: The early life history of fish (ed. Blaxter, J.H.S.), pp. 465-485. Springer-Verlag, Berlin.

Deplano, M., Diaz, J.P., Connes, R., Kentouri-Divanach, M. and Cavalier, F. 1991. Appearance of lipid absorption capacities in larvae of the sea bass *Dicentrarchus* labrax during transition to the exotrophic phase. Marine Biology 108: 361-371.

Divanach, P. and Kentouri, M. 2000. Hatchery techniques for specific diversification in Mediterranean finfish larviculture. In: Cahier Options Mediterranennes, Vol. 47, Mediterranean Marine Aquaculture Finfish Species Diversification, C.I.H.E.A.M. (Ed. Basurco, B.) pp.75-87.Zaragoza, Spain.

Elbal, M.T., García Hernández, M.P., Lozano, M.T. and Agulleiro, B. 2004. Development of the digestive tract of gilthead seabream (*Sparus aurata* L.). Light and electron microscope studies. Aquaculture 234: 215-238.

F.A.O. (Food and Agriculture Organization of the United Nations) 2016. The state of world fisheries and aquaculture. Fisheries and Aquaculture Department, Rome, Italy. 197pp.

F.A.O. (Food and Agriculture Organization of the United Nations) 2010. The state of world fisheries and aquaculture. Fisheries and Aquaculture Department, Rome, Italy. 204pp.

Fuiman, L.A. 1983. Growth gradients in fish larvae. Journal of Fish Biology 23:117-123.

Galaviz, M.A., García-Ortega, A., Gisbert, E., López, L.M. and García Gasca, A. 2012. Expression and activity of trypsin and pepsin during larval development of the spotted rose snapper *Lutjanus guttatus*. Comparative Biochemistry and Physiology 161B:9–16

Gatesoupe, F.J. 1999. The use of probiotics in aquaculture. Aquaculture 180: 147-165.

Gisbert, E. 1999. Early development and allometric growth patterns in Siberian sturgeon and their ecological significance. Journal of Fish Biology 54:852-862.

Gisbert, E. and Doroshov, S.I. 2006. Allometric growth in green sturgeon larvae. Journal of Applied Ichthyology 22 (Suppl.1):202-207.

Gisbert, E., Merino, G., Muguet, J.B., Bush, D., Piedrahita, R.H. and Conklin, D.E. 2002. Morphological development and allometric growth patterns in hatchery-reared California halibut larvae. Journal of Fish Biology. 61: 1217-1229.

Gisbert, E., Piedrahita, R.H. and Conklin, D.E. 2004. Ontogenetic development of the digestive system in California halibut (*Paralichthys californicus*) with notes on feeding practices. Aquaculture 232: 455-470.

Gomez-Gil, B., Roque, A. and Turnbull, J. F. 2000. The use and selection of probiotic bacteria for use in the culture of larval aquatic organisms. Aquaculture 191: 259–270

Gopalan, C., Ramasastri, B.V. and Balasubramanian, S.C. 2004. Nutritive value of Indian Foods. National Institute of Nutrition (NIN). Indian Council of Medical Research. pp: 59-67. Hyderabad,India.

Gouveia, L. and Rema, P.2005. Effect of microalgal biomass concentration and temperature on ornamental goldfish (*Carassius auratus*) skin pigmentation. Aquaculture Nutrition 11: 19–23.

Govoni, J.J., Boehlert, G.W. and Watanabe Y. 1986.The physiology of digestive in fish larvae. Environmental Biology of Fishes 16: 59-77.

Guidelines for cage culture in inland open water bodies of India. 2016. National fisheries development board, Hyderabad, India.14pp.

Hamlin, H.J., Hunt von Herbing, I. and Kling, L.J. 2000. Histological and morphological evaluations of the digestive tract and associated organs of haddock throughout post-hatching ontogeny. Journal of Fish Biology 57: 716-732.

Heming, T.A. and Buddington, R.K. 1988. Yolk sac absorption in embryonic and larval fishes. In: Fish physiology, volume 11 (eds. Hoar, W.S. and Randall, D.J.), pp. 407-446. Academic Press, New York.

Hickman, R. W. 1979. The future of aquaculture in the Australia and New Zealand region. New Zealand Agricultural Science 13 (4): 171-176.

Hoehne-Reitan, K. and Kjørsvik, E. 2004. Functional development of the liver and exocrine pancreas in teleost fish. In: The Development of Form and Function in Fishes and the Question of Larval Adaptation (ed. Govoni, J.J.) American Fisheries Society Symposium. pp: 9-36. Bethseda, MD.

Hora, S.L. 1951. Fish culture in rice fields. Current Science 20 (7): 171–173.

Hossain,Q.Z., Hossain, M.A. and Parween, S. 2006. Breeding biology, captive breeding and fry nursing of humped featherback (*Notopterus chitala*, Hamilton-Buchanan, 1822). Ecoprint 13:41-47.

Hummel, B.C.W. 1959. A modified spectrophotometric determination of chymotrypsin, trypsin and thrombin. Canadian Journal of Biochemistry and Physiology 37: 1393–1399.

IUCN, 2014. IUCN Red List of Threatened Species. Version 2014.1.

Jahan, D.A., Rasid, J., Khan, Md. M. and Mahmud,Y.2015. Early embryonic and larval development of threatened humped featherback, *Chitala chitala* (Hamilton). Trends in fisheries Science 4 (2)1-7.

Jakhar, J.K., Pal, A.K., Devivaraprasad R.A., Sahu, N.P. Venkateshwarlu, G. and Vardia, H.K. 2012. Fatty acid Composition of Some selected Indian Fishes. African Journal of Basic and Applied Sciences 4 (5): 155-160

Jhingran,V.G., 1991. Fish and Fisheries of India, third ed. Hindusthan Publishing Corporation (India), Delhi, pp. 1–727.

Job, T.J., David, A. and Das, K.N. 1955. Fish and fisheries of the Mahanadi in relation to the Hirakund dam. Indian Journal of Fisheries 2(1): 1-40.

Johnston, I.A. and Hall, T.E. 2004. Mechanisms of muscle development and responses to temperature change in fish larvae. In: The development of form and function in fishes and the question of larval adaptation.(Ed. Govoni, J.J.) pp. 85-116. American Fisheries Society, Symposium 40, Bethesda, Maryland.

Kammerer, C.H., Grande, L. and Westneat, M.W. 2005. Comparative and developmental functional morphology of the jaws of living and fossil gars (Actinopterygii: Lepisosteidae). Journal of Morphology 263:1-15.

Kendall, A.W., Ahlstrom, E.H. and Moser, H.G.1984. Early life history stages of fishes and their characters. In: Ontogeny and systematics of fishes (Eds. Moser, H.G., Richards, W.J., Cohen, D.M., Fahay, M.P.,Kendall, A.W. and Richardson, S.L. American Society of Ichthyologists and Herpetologists, Special Publication No. 1, pp. 11-22. Allen Press Inc, Lawrence, Kansas, U.S.A.

Kesarcodi-Watson, A., Kaspar, H., Lategan, M.J. and Gibson, L. 2008. Probiotics in aquaculture: The need, principles and mechanisms of action and screening processes. Aquaculture 274: 1-14.

Kristjansson, B.K. 2005. Rapid morphological changes in threespine stickleback, *Gasterosteus aculeatus*, in freshwater. Environmental Biology of Fishes 74: 357-363.

Kohinoor, A.H.M., Jahan, D.A., Khan, M.M., Islam, M.S. and Hussain, M.G. Reproductive biology of feather back, chital (*Notopterus chitala*, Ham.) cultured in a pond of Bangladesh. International Journal of Agricultural Research in Innovation and Technology 2 (1): 26-31.

Kolmann, M.A. and Huber, D.2009. Scaling of feeding biomechanics in the horn shark *Heterodontus francisci*: ontogenetic constraints on durophagy. Zoology 112:351-361.

Koumoundouros, G., Divanach, P. and Kentouri, M.1999. Ontogeny and allometric plasticity of *Dentex dentex* (Osteichthyes: Sparidae) in rearing conditions. Marine Biology 135:561-572.

Kumar, V., Sahu, N.P., Pal, A.K., Kumar, S., Sinha, A.K., Ranjan, J. and Baruah, K. 2010. Modulation of key enzymes of glycolysis, gluconeogenesis, amino acid catabolism, and TCA cycle of the tropical freshwater fish *Labeo rohita* fed gelatinized and nongelatinized starch diet. Fish Physiology and Biochemistry 36: 491–499.

Kundu, N. and Homechaudhuri, S.2014.Photoperiod as Cues for Determining Optimal Foraging in Developmental Stages in *Chitala chitala* (Hamilton, 1822). Proceedings of Zoological Society 67: 167-174.

Lowry, O. H., Rosebrough, N. J., Farr, A. L. and Randall, R. J. 1951. Protein measurement with the folin phenol reagent. Journal of Bioligal Chemistry 193: 265-275.

Mandal, A., Mohindra, V., Singh, R.K., Punia, P., Singh, A.K. and Lal, K.K. 2012. Mitochondrial DNA variation in natural populations of endangered Indian featherback fish, *Chitala chitala*. Molecular Biology Report 39:1765-75.

Martinez, I., Moyano, F.J., Fernandez-Diaz, C. and Yufera, M. 1999. Digestive enzyme activity during larval development of the Senegal sole (*Solea senegalensis*). Fish Physiology and Biochemistry 21: 317–323.

Menon,A.G.K.1974. A checklist he fishes o the himalayan and the Indo gangetic Plains.Inland Fisheries society India,Barrackpore,136pp.

Mgaya, Y.D. and Mercer, J.P. 1995. The effects of size grading and stocking density on growth performance of juvenile abalone, *Haliotis tuberculata* Linn. Aquaculture 136: 297-312.

Micale, V., Garaffo, M., Genovese, L., Spedicato, M.T. and Muglia, U. 2006. The ontogeny of the alimentary tract during larval development in common pandora, *Pagellus erythrinus*, L. Aquaculture 251:354-465.

Misra, K.S. 1959. An aid to the identification of commercial fishes of India and Pakistan. Records of Indian Museum 57 (1-4):1-320.

Mitra, A. and Mukhopadhyay, P.K.2012. *Chitala chitala* an ideal species for aquaculture diversification. Agrovet Buzz 5(3):63-67.

Mitra, A., Mukhopadhyay, P.K. and Homechaudhuri, S. 2014a. Understanding probiotic potentials of Bacillus Bacterial population isolated from *Chitala chitala* (osteoglossiformes; notopteridae) by comparing the enzyme activity in vitro. International Journal of Pure and Applied Zoology 2:120-127.

Mitra, A., Mukhopadhyay, P.K. and Homechaudhuri, S. 2014 b. Allometric growth pattern during early developmental stages of Featherback, *Chitala chitala* (Hamilton 1822) (Osteoglossiformes: Notopteridae). Asian Fisheries Science 27: 260–273.

Mitra, A., Mukhopadhyay, P.K. and Homechaudhuri, S. 2015a. Histomorphological study of gut developmental pattern in the early life history stages of featherback, *Chitala chitala* (Hamilton). Archive of Polish Fisheries 23: 25–35.

Mitra, A.2015b. The Profile of Selected Digestive Enzymes And Gut Histomorphology Through The Developmental Stages of Featherback, *Chitala chitala* (Hamilton, 1822), With A Probiotic Approach For Nutritional Supplementation. Ph.D. thesis dissertation. University of Calcutta. Department of Zoology.

Mitra, A., Mukhopadhyay, P.K. and Homechaudhuri, S. 2016a. Profile of digestive enzymes activity during early development of featherback *Chitala chitala* (Hamilton, 1822). Proceedings of Zoological Society, doi:10.1007/s12595-016-0169-8.

Mitra, A. 2016b.Chital mach kichu jyatabya bishoy.Matysa Sambad 1(3):34-37.

Mittermeier, R.A. and Mittermeier, C.G. 1997. Megadiversity: Earth's Biologically Wealthiest Nation. In: Global freshwater Biodiversity (Eds. Mc.Allister, D.E., Hamilton, A.L. and Harvery, B.). Vol.11. pp: 1- 140.Sea Wind, Cemex, Mexico City.

Mohanta, K. N., Subramanian, S., Komarpanti, N. and Saurabh, S.2008. Alternate carp species for diversification in fresh water aquaculture in India. Aquaculture Asia 13(1): 11-14.

Mohanty, S.K.1975. Some observations on the physico-chemical features of the outer channel of the Chilika lake during 1971-73. Bulletin of Department of Marine Science 7(1): 69-89.

Mohapatra, S., Chakraborty, T., Prusty, A. K., Das, P., Paniprasad, K. and Mohanta, K. N. 2012. Use of different microbial probiotics in the diet of rohu, *Labeo rohita* fingerlings: effects on growth, nutrient digestibility and retention, digestive enzyme activities and intestinal microflora. Aquaculture Nutrition 18: 1– 11.

Morote, E., Olivar, M.P., Pankhurst, P.M., Villate, F. and Uriarte, I. 2008. Trophic ecology of bullet tuna *Auxis rochei* larvae and ontogeny of feeding-related organs. Marine Ecological Progress Series 353:243-254.

Moutopoulos, D.K. and Stergiou, K.I. 2002. Length-weight and length-length relationship of fish species from the Aegean Sea (Greece). Journal of Applied Ichthyology 18: 200-203.

Moyano, F.J., Diaz, M., Alarcon, F.J. and Sarasquete, M.C.1996. Characterization of digestive enzyme activity during larval develpoment of gilthead seabream *Sparus aurata*. Fish Physiology and Biochemistry 15: 121–130.

Müller, U.K. and van Leeuwen, J.L. 2006. Undulatory fish swimming: from muscles to flow. Fish and Fisheries 7:84- 103.

Munilla-Moran, R., Stark, J.R. and Barbour, A. 1990. The role of exogenous enzymes in digestion in cultured turbot larvae *Scophtalmus maximus*. Aquaculture 88: 337–350.

Murata, Y., Tamura, M., Aita, Y., Fujimura, K., Murakami, Y., Okabe, M., Okada, N. and Tanaka, M. 2010. Allometric growth of the trunk leads to the rostral shift of the pelvic fin in teleost fishes. Developmental Biology 347:236-245.

Near, T. J., Eytan, R. I., Dornburg, A., Kuhn, K. L., Moore, J. A., Davis, M. P. and Smith, W. L. 2012. Resolution of ray-finned fish phylogeny and timing of diversification. Proceedings of the National Academy of Sciences 109(34), 13698-13703.

Ortíz-Delgado, J.B., Darias, M.J., Cañavate, J.P., Yúfera, M. and Sarasquete, C. 2003. Organogenesis of the digestive tract in the white seabream, *Diplodus sargus*. Histological and histochemical approaches. Histology and Histopathology 18: 1141 – 1154

Ortíz-Galindo, J.L., Peña, R., Perezgomez-Alvarez, L. and Castro-Aguirre, J.L. 2000. Desarrollo osteológico de la cabrilla arenera Paralabrax maculatofasciatus (Steindachner, 1868) (Percoidei: Serranidae). Memorias VII Congreso Nacional Ictiología. México, Distrito Federal

Osse, J.W. and Boogaart, J.G.M. 2004. Allometric growth in fish larvae: timing and function. American Fisheries Society Symposium 40:167-194.

Osse, J.W. and Boogaart, J.G.M.1999. Dynamic morphology of fish larvae, structural implications of friction forces in swimming, feeding and ventilation. Journal of Fish Biology 55:156-174.

Ozaki, N. 1965. Some observation on the fine structure of the intestinal epithelium in some marine teleosts. Archive of Histology Japan 26: 23-38.

Palumbo, S.A.1985. Starch-ampicillin agar for the quantitative detection of *Aeromonas hydrophila*. Applied and Environmental Microbiology 50: 1027–1030.

Papandroulakis, N., Kentouri, M., Maingot, E. and Divanach, P. 2004. Mesocosm: a reliable technology for larval rearing of *Diplodus puntazzo* and *Diplodus sargus sargus*. Aquaculture International 12: 345-355.

Parmeshwaran, S. and Sinha, M. 1966. Observation on the biology of featherback *Notopterus notopterus* (Pallas). Indian Journal of Fishes 13 (1&2): 232-250.

Parra, G. and Yúfera, M. 2001. Comparative energetics during early development of two marine fish species, *Solea senegalensis* (Kaup) and *Sparus aurata* (L.). Journal of Experimental Biology 204:2175-2183.

Pearse, A.G.E. 1985.Histochemistry. Theoretical and applied. Vol. 2 Analytic Technology.

Peighambardousta, S.H., Tafti, A.G. and Hesari, J. 2011.Application of spray drying for preservation of lactic acid starter cultures: a review. Trends in Food Science and Technology 22: 215–224.

Pillay, T. V. R. 1990. Aquaculture: Principles and Practices. Fishing New Books, Blackwell Science, Oxford, England, 575pp .

Polling, L. 1993. Ecology of the Smaller Fish Species of a Subtropical Impoundment in the Eastern Transvaal Lowveld. M.Sc. Thesis, Univ. of the North, South Africa.

Pond, M.J., Stone, D.M. and Alderman, D.J. 2006. Comparison of conventional and molecular techniques to investigate the intestinal microflora of rainbow trout (*Oncorhynchus mykiss*). Aquaculture 261: 194-203.

Ponniah, A.G. and Sarkar, U.K. 2000. Overview of fish biodiversity of north east India. In: Fish biodiversity of north east India. (Eds. Ponniah, A.G. and Sarkar,U.K.) NATP Publ. pp.1-102. NBFGR, Lucknow, India.

Pradhan, P.K., Jena, J.K., Mitra, G., Sood, N. and Gisbert, E.2012. Ontogeny of the digestive tract in butter catfish *Ompok bimaculatus* (Bloch) larvae. Fish Physiology and Biochemistry 38: 1601–1617.

Prasad, B. and Mookerji, D.O.1930. On the Fishes of the Manchar lake (Sindh). Journal Bombay Natural History Society 34:164-169.

Qureshi, T.A. and Qureshi, N.A. 1983. Indian Fishes Publishers Brij Brothers, Sultania Road Bhopal (M.P.). 5-209.

Radheyshyam and Sarangi, N. 2005. Breeding and egg incubation of *Notopterus Chitala* (Hamilton) in captivity. Journal of Inland Fisheries Society of India 37(2):8-14.

Rahimibashar, M.R., Alipour, V.P., Hamidi, and Hakimi, B. 2012. Biometric characteristics, diet and gonad index of Lizardfish (Sauridatumbil, Bloch 1795) in North of the Persian Gulf. World Journal of Fish and Marine Sciences 4: 01 06.

Rahman, A.K.A. 1989. Freshwater Fishes of Bangladesh. Zoological Society of Bangladesh, 364pp.

Refstie, S., Landsverk, T., Bakke-McKellep, A.M., Ringø, E., Sundby, A., Shearer, K.D. and Krogdahl, A. 2006. Digestive capacity, intestinal morphology, and microflora of 1-year and 2-year old Atlantic cod (*Gadus morhua*) fed standard or bioprocessed soybean meal. Aquaculture 261: 269-284.

Rick, W. 1974a. Trypsin. In: Methods of enzymatic analysis, vol. 2, ed. H.U. Bergmeyer, 1021–1024. New York: Academic Press.

Rick, W. 1974b. Chymotrypsin. In: Methods of enzymatic analysis, vol. 2, ed. H.U. Bergmeyer, 1009–1012. New York: Academic Press.

Roberts, T.R. 1992. Systematic revision of the old world freshwater fish family Notopteridae. Ichthyological Exploration of Freshwater 2(4): 361-383

Ross, L. G., Palacios, C. M. and Morales, E. 2008. Developing native fish species for aquaculture: the interactive demand for biodiversity, sustainable aquaculture and livelihoods. Aquaculture Research 39: 675-683.

Russo, T., Costa, C. and Cataudella, S. 2007. Correspondence between shape and feeding habit changes throughout ontogeny of gilthead sea bream *Sparus aurata* L. (1758). Journal of Fish Biology 71:629-656.

Sadhale, N. and Nene, Y.L. 2005. On fish in Manasollasa (c. 1131 AD). Asian Agri-History 9:177–199.

Sánchez-Amaya, M.I., Ortíz-Delgado, J.B., García-López, Á. , Cárdenas, S. and Sarasquete, C. 2007. Larval ontogeny of red banded seabream *Pagrus auriga* (Valenciennes, 1843) with special reference to the digestive system. A histological and histochemical approach. Aquaculture 263: 259-279.

Sarasquete, C., Gisbert, E., Ribeiro, L., Vieira, L. and Dinis, M.T. 2001. Glycoconjugates in epidermal, branchial and digestive mucous cells and gastric glands of gilthead sea bream, *Sparus aurata*, Senegal sole, *Solea senegalensis* and Siberian sturgeon, *Acipenser baeri* development . European Journal of Histochemistry 45:267-278.

Sargent, J.R., Henderson, J.R. and Tocher, D.R., 1989. The lipids. In: Fish Nutrition (Ed. Halver, J.E.), 2nd edn. pp. 153–218.Academic Press, New York.

Sarkar, U.K., Lakra, W.S., Deepak, P.K., Negi, R.S., Paul, S.K. and Shrivastava, A. 2006a. Performance of different types of diets on experimental larval rearing of endangered *Chitala chitala* (Ham.) in recirculatory system. Aquaculture 261:141-150.

Sarkar, U.K., Deepak, P.K., Negi, R.S., Singh S.P. and Kapoor, D. 2006b. Captive breeding of endangered fish *Chitala chitala* (Hamilton- Buchanan) for species conservation and sustainable utilization. Biodiversity and Conserversation 15:3579-3589.

Sarkar, U. K., Deepak, P.K., Negi, R.S., Qureshi, T.A. and Lakra, W.S. 2007. Efficacy of different types of live and non-conventional diets on endangered clown knife fish *Chitala chitala* (Hamilton) during its early life stages (ELS). Aquaculture Research 38: 1404- 1410.

Sarkar, U.K., Negi, R.S., Deepak, P.K., Lakra, W.S. and Paul, S.K. 2008. Biological parameters of endangered *Chitala chitala* (Osteoglossiformes: Notopteridae) from some Indian rivers. Fisheries Research 90: 170- 177.

Sarkar, U.K., Deepak, P.K. and Lakra, W.S. 2009a. Stock identification of endangered clown knife fish *Chitala chitala* (Hamilton Buchanan, 1822) from Indian rivers inferred by morphological attributes. Electronic Journal of Ichthyology 2: 59-75.

Sarkar, U.K. and Deepak, P.K. 2009b. The diet of clown knife fish *Chitala chitala* (Hamilton Buchanan) an endangered notopterid from different wild population, India. Electronic Journal of Ichthyology 1: 11-20.

Sarkar, U. K., Deepak, P. K. and Negi, R. S. 2009c. Length–weight relationship of clown knifefish *Chitala chitala* (Hamilton 1822) from the River Ganga basin, India. Journal of Applied Ichthyology 25: 232–233.

Sehgal, K.L. 1973. Fisheries survey of Himachal Pradesh and some adjacent areas with special reference to trout, Mahseer and allied species. Journal of Bombay Natural Society 70(3): 458–471.

Shankar, D. S. and Kulkarni, R. S. 2005. Somatic condition of the fish, *Notopterus notopterus* (pallas) during phases of the reproductive cycle. Journal of Environmental Biology 26(1): 49-53

Shaw, G.E. and Shebbeare, E.O. 1937. The fishes of Northern Bengal. Journal of Royal Asiatic Society of Bengal (Science): 137 + 6 pls.

Singh, S.B., Dey, R.K., Reddy, P.V.G.K. and Mishra, B.K.1980. Some observation on breeding, growth, and fecundity of *Notopterus chitala*. Journal of Inland Fisheries Society of India 12 (2): 13-17.

Skrodenyte-Arba iauskiene, V., Sruoga, A., Butkauskas, D. and Skrupskelis, K. 2008. Phylogenetic analysis of intestinal bacteria of freshwater salmon Salmo salar and sea trout Salmo trutta trutta and diet. Fisheries Science 74: 1307-1314.

Sonnenschein, A.L., Losick, R. and Hoch, J.A., 1993. *Bacillus subtilis* and Others Gram-Positive Bacteria: Biochemistry, Physiology and Molecular Genetics. American Society for Microbiology, Washington, DC, 987pp.

Southwell, T. and Prasad, B. 1918. Notes from Bengal fisheries laboratory. No.5. Parasites from Indian fishes with a note on the Carcinoma in the climbing perch. Records of the Indian Museum 15:341-355.

Spanggaard, B., Huber, I., Nielsen, J., Nielsen, T., Appel, K.F. and Gram, L. 2000. The microflora of rainbow trout intestine: a comparison of traditional and molecular identification. Aquaculture 182: 1-15.

Spyridakis, P., Metailler, R., Gabaudan, J. and Riaza, A. 1989. Studies on nutrient digestibility in European sea bass (Dicentrarchus labrax). Methodological aspects concerning faeces collection. Aquaculture 77:61-70.

Sreekanth, G.B., Tincy,V., ,Mishal,P., Sandeep, K.P. and Praveen, K.P. 2013.Food security in India; Is Aquaculture a solution in the offing? International journal of science and research 4(3):553-560.

Sugita, H., Kawasaki, J. and Deguchi, Y. 1997. Production of amylase by intestinal microflora in cultured freshwater fish. Letter of Applied Microbiology 24:105–108.

Suzer, G., Firat, K. and Saka, S. 2006. Ontogenic development of the digestive enzymes in common pandora, *Pagellus erythrinus,* L. larvae. Aquaculture Research 37(15): 1565-1571.

Suzer, G., Aktülün, S., Çoban, D., Kamaci, H.O., Saka, Ş., Fırat, K. and Alpbaz, A. 2007. Digestive enzyme activities in larvae of sharpsnout sea bream (*Diplodus puntazzo*). Comparative Biochemistry and Physiology 148A: 470- 477.61

Talwar, P.K. and Jhingran, A.G. 1991. Inland fishes of India and adjacent countries. vol 1. A.A. Balkema, Rotterdam. 541 p.

Tocher, D.R. and Sargent, J.R., 1984. Analyses of lipid and fatty acids in ripe roes of some northwest European marine fish. Lipids 19: 492–499.

Uscanga-Martínez, A., Velázquez-Velázquez, E, Rodríguez-Valencia, W. and Gómez-Gómez, M.A.2012. "Estudio reproductivo para el cultivo de mojarra nativa Cichlasoma trimaculatum" Libro de resúmenes del XIII congreso nacional de ictiología y 1er simposio latinoamericano de ictiología. Universidad de Ciencias y Artes de Chiapas, México, 28 de octubre al 02 de noviembre 2012

van Snik, G.M.J., van den Boogaart, J.G.M. and Osse, J.W.M. 1997. Larval growth patterns in *Cyprinus carpio* and *Clarias gariepinus* with attention to finfold. Journal of Fish Biology 50:1339-1352.

Verreth, J.A., Torrelle, E., Spazier, E. and Sluiszen, H.W. 1992. The development of a functional digestive system in the African catfish, *Clarias garipinus.* Journal of World Aquaculture Society 23: 286-298.

Verschuere, L., Rombaut, G., Sorgeloos, P. and Verstraete, W. 2000. Probiotic bacteria as biological control agents in aquaculture. Microbiology and Molecular Biology Review 64: 655-671.

Vicentini, R. N. and Araujo, F. G. 2003. Sex ratio and size structure of *Micropogonias furnieri* (Desmarest, 1823) (Perciformes, Sciaenidae) in Sepetiba bay, Rio de Janeiro, Brazil. Brazilian Journal of Biology 3: 559-566.

Walker, J.A. 2004. Kinematics and performance of maneuvering control surfaces in teleost fishes. IEEE J. Oceanic Engineering 3:572-584.

Welsch, U. and Storch, V. 1976. Comparative animal cytology and histology. Sidgwick and Jackson Limited, London. 235pp.

Williams, P.J., Brown, J.A., Gotceitas, V. and Pepin, P. 1996. Development changes in escape response performance of five species of marine larval fish. Canadian Journal of Fish Aquacultural Science 53:1246-1253.

Yang, D.Q., Chen, F. and Liu, B.T. 1997. Preliminary study on the food composition of mud eel, *Monopterus albus.* Acta Hydrobiologia Sinica 21: 24-30.

Yang, G., Bao, B., Peatman, E., Li, H., Huang, L. and Ren, D. 2007. Analysis of the composition of the bacterial community in puffer fish *Takifugu obscurus.* Aquaculture 262: 183-191.

Yúfera, M., Sarasquete, M.C. and Fernández-Díaz, C. 1996. Testing protein-walled microcapsules for the rearing of first-feeding gilthead sea bream (*Sparus aurata* L.) larvae. Marine and Freshwater Research 47: 211-216.

Zambonino-Infante, J.L. and Cahu, C.L.1994a. Influence of diet on pepsin and some pancreatic enzymes in sea bass (*Dicentrarchus labrax*) larvae. Comparative Biochemistry and Physiology 109A: 209-212.

Zambonino-Infante, J.L. and Cahu, C.L.1994b. Development and response to a diet change of some digestive enzymes in sea bass (*Dicentrarchus labrax*) larvae. Fish Physiology and Biochemistry 12: 399-408.

Zambonino-Infante, J.L., Cahu, C.L., Peres, A., Quazuguel, P. and Le Gall, M.M.1996. Sea bass *Dicentrarchus labrax* larvae fed different *Artemia* rations: growth, pancreas enzymatic response and development of digestive functions. Aquaculture 139:129-138.

Zambonino-Infante, J.L., Gisbert, E., Sarasquete, C., Navarro, I., Gutiérrez, J. and Cahu C.L. 2008. Ontogeny and physiology of the digestive system of marine fish larvae. In: Feeding and digestive functions in fishes (Eds. Cyrino, J.E.P., Bureau, D. and Kapoor, B.G.). Science Publishers, Enfield, New Hampshire, USA.

Zambonino-Infante, J.L. and Cahu, C.L. 2001. Ontogeny of the digestive tract of marine fish larvae. Comparative Biochemistry and Physiology 130 C: 477-487.

Ziaei-Nejad, S., Rezaei, M.H., Takami, G.A., Lovett, D.L., Mirvaghefi A.R. and Shakouri, M. 2006. The effect of *Bacillus* spp. bacteria used as probiotics on digestive enzyme activity, survival and growth in the Indian white shrimp *Fenneropenaeus indicus*. Aquaculture 252(2-4): 516-524.

www.ingramcontent.com/pod-product-compliance
Ingram Content Group UK Ltd.
Pitfield, Milton Keynes, MK11 3LW, UK
UKHW021011290726
14059UKWH00001BA/81

9 789390 384051